BIOPHILIC CONNECTIONS AND ENVIRONMENTAL ENCOUNTERS IN THE URBAN AGE

Biophilic Connections and Environmental Encounters in the Urban Age takes a multi-disciplinary approach, drawing on the authors' wide range of experience, to provide a greater understanding of the different dimensions of environmental engagement. It considers the ways that we interact with our environments, presenting a comprehensive account of how people negotiate and use the urban landscape.

Set within current debates concerning urban futures, societal issues, sustainable cities, health and well-being, the book explores our innate need for contact with the natural world through biophilic design thinking to expand our knowledge base and promote a wider understanding of the importance of these interactions on our collective well-being. It responds to questions such as, what are the urban qualities that support our well-being? As an urbanised society, what are the environmental determinants that promote healthy and satisfying lifestyles? Beginning with an overview of concepts relating to biophilia and environmental engagement, it moves through current theory and practice, different pathways and their characteristics, before presenting real-world examples and applications through illustrated case studies in different world situations.

With a particular focus on the experience of individuals, the book is essential reading for students, researchers, and practitioners of architecture, landscape architecture, urban planning, design and health sciences, interested in the future of our cities and the importance of green spaces.

Richard Coles is Emeritus Professor of Landscape in the Birmingham School of Architecture and Design, Birmingham City University, UK. His research involves understanding the nature of environmental interaction and the development of environments that are supportive of the needs of users as well as broader issues relating to urban greening. His initial training in biological sciences, and his work within a design school of architecture, allows him to adopt a unique stance to investigation. During his academic career, he has received a number of grant awards from the UK Research Councils and EU funding as well as supervising numerous research studies.

Sandra Costa is a Senior Lecturer in Landscape Architecture at Birmingham School of Architecture and Design, Birmingham City University, UK. Her research examines the nature of person-place interactions, exploring the choreographies of the experience and how individuals negotiate well-being in the landscape. Sandra's research informs and is informed by teaching design studio, live projects and transdisciplinary collaborations, where she further investigates how we can design environments for human well-being whilst responding to and interacting with a changing climate.

BIOPHILIC CONNECTIONS AND ENVIRONMENTAL ENCOUNTERS IN THE URBAN AGE

Frameworks and Interdisciplinary Practice in the Built Environment

Richard Coles and Sandra Costa

Routledge
Taylor & Francis Group
LONDON AND NEW YORK
earthscan
from Routledge

Designed cover image: © Susan Coles - Sydney city centre viewed from the Botanic Gardens

First published 2023
by Routledge
4 Park Square, Milton Park, Abingdon, Oxon OX14 4RN

and by Routledge
605 Third Avenue, New York, NY 10158

Routledge is an imprint of the Taylor & Francis Group, an informa business

British Library Cataloguing-in-Publication Data
A catalogue record for this book is available from the British Library

ISBN: 978-0-367-56875-7 (hbk)
ISBN: 978-0-367-56874-0 (pbk)
ISBN: 978-1-003-09975-8 (ebk)

DOI: 10.4324/9781003099758

Typeset in Bembo
by Apex CoVantage, LLC

CONTENTS

FIGURES

BOXES

ACKNOWLEDGMENTS

The authors would like to express their special thanks to those who have contributed to the Case Studies chapter with insights and contents:

Annette Vennard (Forest Schools, Our Lady Queen of Peace, Primary School)

Arup, UK; Felicitas zu Dohna (Senior Consultant, Foresight), Daniel Imade, Isabel Heinemann; Darren Soh; Thomas Graham

Lin Bolt (Public Relations, WOHA Architects)

Dr Anuj Jain (Principal Ecologist & Biomimic at bioSEA)

Tim South (LDA Design)

Matthew Wood (SiteWorkshop, Seattle)

Odace Évènements (Montreal)

PCA-STREAM (Paris)

Lee Good

Susan Coles

Cllr. Andy Taylor (Warndon Parish Council)

Dr. Laura Mumaw (Research Associate, School of Global, Urban and Social Studies RMIT University).

INTRODUCTION

Connecting with the Landscape

Over the years we have collected data from many landscape situations regarding how people access and interact with the environment. In particular, we have focused on recording the experiences of the individual, examining the types of situations people prefer, including their feelings as they move through different places and one of the strongest aspects has been how people describe their encounters. Particularly striking is the eloquence of the language: the words used are not those of the professional, they are different, full of emotion and descriptions that showed what different places meant to them, how sophisticated interaction is and how their encounters are of great value to personal lifestyles. People have very clear ideas about the significance of these places to their lives and, when prompted, are able to express them.

Our focus has been on the landscape and especially its natural properties, how nature is contained and featured in the city, but also how it is represented in many diverse forms, ranging from bits of countryside encapsulated by urban development to places newly created as part of comprehensive landscape strategies, newly planted woodlands, grasslands and parks, landscapes established to create 'natural' infrastructure; it was the connections that people demonstrated with these spaces, right down to the street tree outside their house, that proved so fascinating. These were places and situations that offered opportunities for 'chilling out', prompted positive reflections on life, refreshed a sense of identity and self-worth, contained memories and facilitated recollections of loved ones (Coles & Bussey, 2000, 1995). For the local population, their usage was a positive aspect of urban living which provided a link to the natural world and nature in all its forms; these features were not an 'add on', but were significant as part of a personal life strategy related to individual needs.

DOI: 10.4324/9781003099758-1

The importance of such 'natural' spaces within an urban complex is now well established as having a positive impact on human well-being, especially mental and physical health (Bowler et al., 2010; Frumkin, 2001), but when we started our study, such knowledge was substantially less advanced. Our research approach was characterised by developing user-based methodologies which revealed how people connected to different environments, the different circumstances surrounding their encounters and the different ways that they were expressed. Each of us negotiates the environment according to our own situation, our desires, background, circumstances and requirements, although we may not be overtly aware of the specifics underpinning engagement (Coles, 2014). However, greater awareness of the process on a personal level, for example through the adoption of techniques which 'focus an awareness of the journey' (Costa & Coles, 2018; Calabrese & Dommert, 2019), can be promoted to more fully realise what should be the positive impact (of interaction) and once understood adapted to other situations in the context of urbanism or urban design (Lynch, 1964; Gehl, 2011). Our experience made it clear that there was a need to promote a wider understanding of the field, i.e. to combine our knowledge with other sources in developing a text exploring environmental engagement, the ways that we connect with the environment as an urban-based society and how we respond to different situations. Accordingly, in this book, we focus on the experience of the individual, exploring ideas relating to environmental interaction, particularly the ways and the circumstances that induce positive outcomes and how such connections are, or might be, embedded in the urban fabric. It was also clear from conversations with others, working with students and other researchers, including wider collaboration across different academic disciplines, that such a text would help advance understanding and similarly help inform professional practice.

Urban Living

A range of studies seek to inform us about how people negotiate and use different urban locations, typically aimed at informing strategies for urban development, and take account of how we encounter buildings, places, navigate journeys, experience nature and perceive the landscape. They seek to provide clarity to those responsible for designing and managing the urban realm, for example, aimed at improving the walking experience and the visibility of the city (Lynch, 1964; Gehl, 2011), or the health and well-being of the population (Bowler et al., 2010; Frumkin, 2001). Such studies are set within different disciplinary frameworks and take account of the complexity of modern urban life and the diversity of urban populations, and as such concern the design of the urban fabric in the pursuit of sustainable city forms which provide 'quality environments' (Crookston, 2001; English Partnerships, 2007).

There is concern that modern urban living can have a corrosive effect upon health (Bowler et al., 2010; Frumkin, 2001) unless it includes appropriate contact with the natural world or 'natural elements'. Cities are highly sophisticated yet, despite our technical sophistication, a lack of contact with the natural world has been shown via research and practice to impact negatively upon health and well-being; conversely, appropriate contact with it has been demonstrated to improve physical and mental health (Bowler et al., 2010; Frumkin, 2001; Kaplan & Kaplan, 1982). There is now a substantial commentary regarding the penalties of isolating ourselves from nature, while many of us can recall the extent to which we were free, as children living in urban centres, to roam in local 'wild' areas unhindered (Louv, 2012). These different studies adopt a range of approaches and measures (Choudhry et al., 2015). Increasingly they make reference to well-being (Coles et al., 2019a) and reflect a greater emphasis on well-being as an appropriate measure of human life satisfaction (Huppert & So, 2013), a theme reflected in the OECD's call to '*redefine the growth narrative to put people's well-being at the centre of governments' effort*' (OECD, 2016, p. 12). Typically, such investigations seek to understand how well-being is compromised, supported or might be supported in the planning and design of modern urban environments that encompass a range of disciplines and investigation techniques (Coles et al., 2019).

Working with other disciplines is enlightening; not only does it enable us to make cross-disciplinary enquiry, it also reveals different facets relating to 'person-place' relationships and methodologies that, for example, are more related to arts practice, being capable of further application in community and other public-facing scenarios. These have revealed facets regarding environmental interaction that, if we hadn't had such conversations, would have remained hidden (Coles et al., 2019). Such dialogue has been so beneficial that we specifically provided a platform for such an exchange of ideas through a series of conferences in which we sought to encourage others to apply their expertise in a similar way (Coles & Millman, 2013; 2016; Coles et al., 2019a).

Biophilic and Salutogenic Environments

A range of theoretical perspectives seek to explain human preference for different landscape types, both in urban and rural locations, and we examine these in the following chapters, but as the reference to well-being suggests, there is particular interest regarding those that have the capacity to promote good health, facilitate recovery from mental stress or recuperation from illness. Environments which promote good health and engender healthy behaviours, such as walking, are termed 'salutogenic' (Vaandrager, 2016), but the theoretical positioning of why they have salutogenic properties is, at least in part, explained by reference to the 'Biophilia Hypothesis' (Wilson, 1984). The Biophilia Hypothesis suggests that we have an innate connection (and need to connect) with nature and biological systems; thus, when we experience landscapes that are biased towards the natural, we tend towards a positive health/well-being association. In the city, the terms nature or natural are somewhat flexible, being explored and defined more specifically in relation to the human experience (Fuller & Irvine, 2010; Coles et al., 2013) where contact with the natural world, or elements of nature, can be provided in many different forms or modes, green spaces obviously, but also through the properties of materials, views, light and other basic natural phenomena (Calabrese & Dommert, 2019; Kellert & Calabrese, 2015).

The Biophilia Hypothesis does help in understanding the underlying properties indicated in person-place relationships, and we take an holistic approach in exploring biophilia and 'biophilic design', (Calabrese & Dommert, 2019; Kellert & Calabrese, 2015) where it is important to enquire about the broader nature of 'biophilic connections', how and where they are contained or realised within different elements of the urban fabric (Salingaros, 2015).

The Context of Sustainability

While the specifics of our investigation concern how individuals interact with the environment in urban areas, and the key focus is on biophilic and salutogenic, or otherwise, properties of the urban environment, the broader context of investigation is sustainable development or, more specifically, the sustainable city (www.un.org/sustainabledevelopment/sustainablecities). As we prepared material for this book and reviewed our own ideas and thoughts in the context of other work, it became even clearer than we had first thought that notions surrounding biophilia play out extremely well in the context of wider thinking regarding the sustainable city. In particular, the biological models that typically underpin sustainable interventions, for example tree planting to address heat island effect, urban watershed and sustainable urban drainage schemes, involve the very components that are indicated in creating biophilic environments and biophilic connections. The point is that there are commonalities of approaches, sustainable interventions that relate extremely well to each other; they indicate solutions which are multifunctional and interrelated (for example, see Bibri & Krogstie, 2019).

The sustainable solutions being advocated and adopted are becoming evident in building forms and urban features. They have given rise to sustainable design practices which, in turn, have given rise to an associated aesthetic which clearly includes biophilic considerations. While such ideas concern new development, they also concern the critical approach of retrofitting, i.e. establishing sustainable solutions within existing city infrastructure, and our cities still retain the capacity to realise opportunities. This applies particularly for adopting biophilic approaches and solutions, although the potential often remains unrealised or overlooked. One of the benefits in working across disciplines, and within an educational setting, especially with those who have chosen career paths which involve designing the built environment, is their interest and enthusiasm regarding sustainability and the future of the environment. Students eagerly seek out information regarding sustainable solutions. Increasingly, ideas relating to biophilia and the quality of life form the basis for developing their own design solutions set in the context of human need, world issues and environmental concerns, which clearly underpin their philosophy and rational for their chosen career paths. It is their enthusiasm as much as anything else that helped inspire us to compile this book.

Exploring Environmental Engagement

Based upon this experience, our combined thinking and working across the different knowledge streams relating to environmental engagement, we have put this book together to enable others to consider our approach and ideas, setting them within the broader field. Accordingly, in this text, we explore the ways that people engage with the urban environment, paying particular attention to the biophilic connections that are indicated in positive encounters. This book pursues a broad outlook, taking the lead from ideas being generated across landscape, planning and architectural disciplines, but investigates and advances these on a substantially wider multidisciplinary basis, targeting aspects of research and practice which inform us of person-place relationships but which originate in different fields. These span different aspects of the built environment/well-being paradigm (Coles et al., 2019a) but are typified by addressing the wider agenda relating to designing for and supporting the well-being of individuals, set within the pursuit of sustainable city forms which respond to the needs of society. Ideas are clearly allied to current thinking concerning urban greening, health and environment, but it is the person-place-environment relationship that underpins our investigation.

In examining this relationship, we attempt to provide greater clarity regarding the achievement of salutogenic and, by implication, sustainable environments, exploring the relationship across different sectors of urban activities. We are aware that it is important to take account of everyday activities, but also to investigate different scenarios which place the potential user at the centre of the equation and thus generate greater understanding of how individuals interact and encounter places that at least offer salutogenic properties.

We aim to:

- Explore the parameters relating to environmental interaction in an urban context investigating the different dimensions of engagement, with particular emphasis on the opportunities and factors underpinning the achievement of a positive 'salutogenic' interface.
- Investigate and demonstrate how different connections are embodied in the urban scene, the pathways presented and how these might be experienced and reinforced.
- Examine how positive environmental encounters are expressed by the individual, supported through different pathways and interventions, with particular emphasis on the everyday.
- Explore thinking relating to ideas of biophilia, in particular taking a holistic approach, to explore the attributes of biophilic thinking with the aim of expanding the knowledge base and promoting a wider understanding of environmental engagement.

We have tried to address a wide audience in presenting ideas, initially introducing the various theoretical perspectives to give an overview of the field and provide knowledge which underpins later exploration. This takes account of the fact that ideas have been generated in different circumstances but, as such, they do address different dimensions of engagement. In particular, these differences need to be explored in terms of identifying pathways of engagement, especially as connections and flows only exist in the reality of real-world situations, and these situations are contained within different disciplinary frameworks and niches of user engagement. In addition, in examining the literature, several commentators make reference to a 'silo approach' to investigation, i.e. that there is little crossover between different 'streams' of investigation and that we also need to take better account of how people access or encounter different environments.

Later chapters investigate the specifics of engagement, drawing on new research findings, with exploration of ideas made possible by methodological advances involving unobtrusive mobile technologies, analysing how we move and connect with the landscape. These chapters also explore the emotional aspects of engagement and ideas that are being generated from the fields of human geography, social science, ethnography and in the context of the arts practitioner. The aim of these chapters is to examine the processes in more depth where they are illustrated by various examples.

Accordingly, the reader is led through to deeper study which addresses a range of situations and perspectives to advance ideas. However, there is also a need to examine these in terms of existing practice, or the different circumstances represented in the real world, and we devote a specific chapter to case studies. These are drawn from different areas of operation to present a cross-section of examples.

A concluding chapter brings ideas together and, overall, it is hoped that the information presented will allow and encourage the reader to consider the different aspects of environmental engagement within their professional remits or circumstances. Places are experienced and explored by people on an everyday basis, where their experience affects how they feel and links to an individual's well-being where it is clear that, in appropriate circumstances, this can be enhanced. We would suggest that, by understanding the parameters relating to engagement and the different situations relating to 'everyday encounters', it is possible to identify the circumstances and impacts of different person-place relationships, especially in supporting the development of and management of urban infrastructure and the biophilic interfaces that need to be considered in both built and natural forms.

References

Bibri, S. E., Krogstie, J. (2019). Generating a vision for smart sustainable cities of the future: A scholarly back casting approach, *European Journal of Futures Research* **7**, 5.

Bowler, D. E., Buyung-Ali, L. M., Knight, T. M., Pulin, A. S. (2010). A systematic review of evidence for the added benefits to health of exposures to natural environments, *BMC Public Health* 10, 456.

Calabrese, E. F., Dommert, A. (2019). Biophilia and the practice of biophilic design. In Coles, R. W., Costa, S., Watson, S. (eds). *Pathways to Well-being in Design Examples from the Arts, Humanities and Built Environment*, Chapter 5. Abingdon: Routledge, pp. 97–127.

Choudhry, K., Coles, R. W., Qureshi, S., Ashford, R., Khan, S., Mir, R. (2015). A review of methodologies used in studies investigating human behaviour as a determinant of outcome for exposure to naturalistic and urban environments, *Urban Forestry & Urban Greening* 14, 527–537.

Coles, R. W. (2014). Environmental interaction and engagement – Supporting well-being. In Burton, E., Davies-Cooper, R., Cooper, C. (eds). *Wellbeing and Environment,* Chapter 17. Oxford: Wiley-Blackwell, pp. 445–499.

Coles, R. W., Bussey, S. C. (1995). Recreation patterns in an urban community forest, *Quarterly Journal of Forestry* 85(3).

Coles, R. W., Bussey, S. C. (2000). Urban forest landscapes in the UK, progressing the social agenda, *Landscape and Urban Planning* 52(2–3), 181–188.

Coles, R. W., Costa, S., Watson, S. (2019). Pathways to well-being. In Coles, R. W., Costa, S., Watson, S. (eds). *Pathways to Well-being in Design-Examples from the Arts, Humanities and Built Environment,* Chapter 12. Abingdon: Routledge, pp. 223–238.

Coles, R. W., Costa, S., Watson, S. (2019a). *Pathways to Well-being in Design-Examples from the Arts, Humanities and Built Environment.* Abingdon: Routledge.

Coles, R. W., Millman, Z. (2013). *Landscape, Well-being and Environment.* Abingdon: Routledge.

Coles, R. W., Millman, Z. (2016). *Designs on Well-being.* Birmingham: Article Press.

Coles, R. W., Millman, Z., Flanningan, J. (2013). Everyday environmental encounters, their meaning and importance for the individual, *Urban Ecosytems* 16, 819–839.

Costa, S., Coles, R. (2018). Self-narrated walking: A user-led method to research people's experiences in urban landscapes, *Landscape Research.* https://doi org/0.1080/01426397.2018.1467004

Crookston, M. (2001). *Urban Design-The New Agenda in Practice.* JPLC.org/files/pdf/2001crookston.pdf

English Partnerships. (2007). *Urban Design Compendium 2.* London: English Partnerships.

Frumkin, H. (2001). Beyond toxicity: Human health and the natural environment, *American Journal of Preventive Medicine* 20(3), 234–240.

Fuller, R. A., Irvine, K. A. (2010). Interactions between nature and people in urban environments. In Gaston, K. (ed). *Urban Ecology.* Cambridge: Cambridge University Press.

Gehl, J. (2011). *Life between Buildings.* Washington: Island Press.

Huppert, F., So, T. (2013). Flourishing across Europe: Application of a new conceptual framework for defining well-being, *Social Indicators Research* 110(3), 837–861.

Kaplan, S., Kaplan, R. (1982). *Cognition and Environment: Functioning in an Uncertain World.* New York: Praeger.

Kellert, S., Calabrese, E. (2015). *The Practice of Biophilic Design.* www.biophilic-design.com

Louv, R. (2012). *The Nature Principle: Reconnecting with Life in a Virtual Age.* Chapel Hill: Algonquin Press.

Lynch, K. (1964). *The Image of the City.* Cambridge: Harvard-MIT.

OECD. (2016). *Strategic Orientation of the Secretary General for 2016 and beyond, Meeting of the OECD Council at Ministerial Level Paris 1–2 June 2016.* www.oecd.org/mcm/document/strategic-orientatiions-of-the-secretary-general-2016.pdf

Salingaros, N. A. (2015). *Biophilia and Healing Environments: Healthy Principles for Designing the Built World.* New York: Terrapin Bright Green, LLC.

Vaandrager, L. (2016). The Application of salutogenesis in communities and neighbourhoods, *NCBI Bookshelf.* www.ncbi.nlm.nih.gov/books/NBK435839

Wilson, E. O. (1984). *Biophilia: The Human Bond with Other Species.* Cambridge: Harvard University Press.

1

THE MULTI-DIMENSIONS OF ENVIRONMENTAL ENGAGEMENT

Introduction

The aim of this chapter is to explore some of the theoretical grounding that underpins discussion concerning human perception of the environment, especially regarding the urban environment. Theory, of course, is not developed in isolation from the context in which it was formulated, or indeed the professional niche of the investigation, so it is necessary to draw information from a range of disciplines in order to come to some conclusions about the different dimensions of environmental engagement. This also means that the circumstances surrounding investigation vary according to the remit, e.g. urban designers who are concerned with improving the quality of the urban environment in relation to the experience of urban populations, through to those whose remit mainly lies in understanding human responses, such as philosophers and environmental psychologists, and also those more concerned with aesthetics, or artistic practice. Despite this, although the terminologies used differ, it is possible to discern common themes which concern how people interact with and make sense of the environment.

As identified in the introduction, our focus has been on the experience of the individual, and we continue this approach in examining the literature, although, as the reader will see, there are commonalities of findings which relate to the broader human condition, including the ways that the senses operate and the body is orientated in space. The urban realm is diverse and is the location where people live, work, relax and carry out their everyday activities, but much of the research concerning how we perceive and engage with the environment falls into two fairly distinct camps – the highly designed urban realm which bears little resemblance to the natural world, which includes all the sights, sounds, activities and complexities of modern living, and the natural or close-to-natural environments that occupy substantial parts of urban centres and have a clear affinity with natural ecosystems even

DOI: 10.4324/9781003099758-2

if they are designed. The key difference is that in one system, nature dominates, and in the other, it recedes. This is, of course, a rather simplistic interpretation of the complexity of the urban environment, but it is helpful at this stage to establish this difference. Accordingly, those more affiliated to the natural world are typified by the inherent properties associated with natural systems where natural sights, sounds and cues dominate the senses, and tend to be explained under one set of theories, and those principally concerning the built form are considered by urban theorists.

Current research, especially that undertaken by environmental psychologists (for example, Von Linden et al., 2022; Kaplan, 2011), tends to draw a distinction between the natural world and the built environment regarding the consequences of exposure to each. The findings suggest that the experience of the 'hard built environment' and the associated consequences of urban living is stress inducing, while the experience of the natural 'green' environment is stress relieving. These two extremes thus represent the two ends of a spectrum where researchers suggest that living in the highly developed urban world, where nature is relegated to a minor component, tends towards causing stress; conversely, exposure to natural green spaces offers restoration from this stressed state, a condition known as 'Attention Restoration Theory' or 'ART' (Von Linden et al., 2022; Kaplan, 1995). These findings emphasise the importance of 'accessible natural environments' to human well-being, but those in the discipline of urban design also explore how we interact with built/planned forms and seek to define the properties of designed systems, city structures and spaces which are far removed from natural systems, all with the aim of improving the quality of life, i.e. exploring how we interact with and move through the city, especially the parameters which underpin positive interaction.

Both sets of investigation emphasise that it is the routine activities associated with life, how people move through the city and experience places, that is important, and tend to emphasise walking as the natural pace of encounter, a pace at which our senses are attuned to assimilate information (O'Neill & Hubbard, 2010; Pink et al., 2010). Environmental psychologists compare the walking experience in natural versus built up areas, but urban conglomerations are a diverse mix of typologies ranging from near-natural to no parallel in the natural world, although many of the designed components actually mimic or have their basis in natural systems and are seen in the parks and landscape spaces scattered throughout a city. Indeed, cities are often defined by the splendour of their landscapes, as well as of course the magnificence of their buildings. Bringing information together starts to reveal how people react to different cues and make sense of places within their own remits, and how they explore them and make sense of different situations, including the meaning that such places engender.

The Context of Biophilia

Especially in the context of urban living, biophilia is explored as a human condition regarding an innate need for contact with nature, which has become more obvious as we are divorced from survival activities that involve some form of natural/nature interface, to the extent that we do not just seek out contact with the natural world, we require it to thrive (Wilson, 1984). Biophilia as a concept has gained substantial credibility over the years, and our understanding of the benefits accrued from accessing nature in terms of mental and physical health and well-being have been recorded in many studies (see, for example, Choudhry et al., 2015). A substantial body of research exists regarding the therapeutic properties of accessing green or natural spaces, with work spanning a range of fields including environmental psychology (Hartig et al., 1991) health promotion and design (Calabrese & Dommert, 2018). Indeed, 'green prescribing' is now seen as an integrated aspect of health care (see, for example, Intelligent Health, n.d. and also Barry & Blythe, 2018). Research also indicates the penalties caused by a lack of access to green spaces, what is termed Nature-Deficit Disorder (Louv, 2008), and reflects that contact with the natural world is much reduced in current urban typologies but also includes discussion surrounding what biophilic typologies might embrace (Calabrese & Dommert, 2018).

In the city, 'natural' areas or their 'designed analogues' may be modified to better optimise human accessibility. It is not difficult to recognise their natural origins, or indeed the biophilic potential of such spaces, but how should we define urban nature, particularly in the context of identifying potential situations where biophilic connections might occur? Bearing in mind the diverse ways that nature is presented or experienced in urban areas, it is possible to consider urban nature as a construct, and in the past, we have adopted the idea of 'nature's constants' (Coles, 2014), reflecting the fact that they offer properties which are beyond the immediate realm of human control, and exist in a slightly different 'natural dimension' (e.g. seasonal change, the effects of time and weathering). They demonstrate an apparent constancy and familiarity, especially when considered against the rapid pace of urban change or life events (Coles, 2014; Coles et al., 2013).

Substantial areas of natural, or representatives of natural, vegetation remain or are created as part of urban design, and it is easy to recognise these as constants; they form the green and blue infrastructure expressed in a wide variety of forms. Their close approximation to natural systems underpins their biophilic properties, offering a complete spectrum of positive biophilic interaction and a complex array of sensory attractions, sights, smells, noise, touch, taste and natural rhythms of engagement, including wider environmental connections such as views to the horizon, contact with the weather, the full impact of seasonal change, the night sky and wildlife. But biophilic connections are also indicated in materials which retain their elemental forms (stone, timber, etc) and their natural textures, and show the effects of weathering, thus reflecting the wider qualities of the environment.

Others explore the presence of nature in urbanised regions as a distinct discipline, 'urban ecology', involving those urban ecosystems that are contained within the city, including highly modified systems as well as those specifically created by human activities. Urban ecology is now a highly respected and established discipline, and urban ecosystems are considered worthy of study and definition as valuable systems to both urban dwellers and wildlife. Urban regions thus exhibit ecological features which reflect their diverse characteristics, their histories and economic activities. We now celebrate the movement of wildlife into urban regions and embrace the presence of such and means of incorporating different niches for different species (Gilbert, 2011).

It is useful to regard the urban realm as a human-focused ecosystem which provides different niches for different activities, and these display individual characteristics of organisation. The urban scene can thus be considered as containing a myriad of different sub-systems ranging from hard components, buildings and roads, through to largely natural places, for example, pieces of encapsulated countryside which still reflect the vegetation characteristics of the region, through to those which are more difficult to describe and those specifically designed for human use, such as parks, botanical gardens, horticultural activities and food growing ventures. Together they make up an environment which is dynamic, but is explored in terms of its properties and how people respond to these properties as they encounter different situations and, in particular, find an affinity with them which overall is found to be absorbing and positive.

Moving Through the City: the Theories of Lynch and Gehl

Moving from urban ecology to urban forms, we now consider how cities are perceived exploring the work of the two well-known urbanists: Kevin Lynch, through *The Image of the City* (originally published in 1960) and Jan Gehl, who looked at 'Life Between Buildings' (originally published in 1971).

Lynch – *The Image of the City*

Kevin Lynch specifically examined how city dwellers perceive and understand their urban centres. At the time, Lynch's ideas and approach were set in the context of an absence of theory or experimentation in the matter of how cities are perceived by their inhabitants (i.e. as opposed to professional designers) based on objective criteria and dealing with visual form at a city/urban scale. Importantly, he takes a user-based approach, interviewing city dwellers regarding their experiences and perceptions of the urban environment to identify the meaning associated with different city elements in relation to the everyday functions of life, a perspective that is particularly appropriate in understanding how people read and navigate their surroundings (Figure 1.1).

Lynch thus derived his ideas from the real experiences of individuals based on USA example cities, to explore what he calls the '"Legibility of the cityscape" . . . By this we mean the ease with which its parts can be recognized', how a mental image of a city is synthesised and how it is a two-way process. To quote Lynch:

> Environmental images are the result of a two-way process between the observer and his environment. The environment suggests distinctions and relations, and the observer, with great adaptability and in the light of his own purposes, selects, organizes, and endows with meaning what he sees . . . the

FIGURE 1.1 Image of the city: Worcester Cathedral and the River Severn. A view which encapsulates the image of the city and how it is read through views and space typologies. This is an enduring view that attracts locals and visitors, containing impressive architecture and nature.

> product both of immediate sensation and of the memory of past experience, and it is used to interpret information and to guide action . . . in particular the people and their activities, are as important as the stationary physical parts. We are not simply observers of this spectacle, but are ourselves a part of it, on the stage with the other participants. Most often, our perception of the city is not sustained, but rather partial, fragmentary, mixed with other concerns. Nearly every sense is in operation, and the image is the composite of them all.
>
> *(Lynch, 1990)*

Lynch emphasises that a flow of information and interpreted meaning exists between the urban experience and the physical manifestation of different city elements which individuals synthesise into a meaningful encounter. In particular, they concern how the urban dweller orientates themselves, navigates the city and how closely it is linked to well-being, '*to enjoy uniqueness and specialization*'. Notably, he recognises that this is based upon a multi-sensory experience, '*nearly every sense is in operation*', although much of his thinking is related to the visual. From his observations and the analysis of interviewees' responses, Lynch was able to define the parameters that relate to the image of a city, how it is read and the component parts which served to define city structure in relation to user experience, perception and navigation. These relate to how sense is made of the complexity, introducing an order of scale and features which make sense, referring to visual strength and impact but also how places are defined by distinctive boundaries and spatial prominence, in which:

- Paths were the predominant city elements by which the city is explored.
- Nodes are the strategic foci into which the observer can enter.
- Districts are the relatively large city areas which the observer can mentally go inside of and which have some common character.
- Edges are the linear elements and act as lateral references, typically boundaries.
- Landmarks are point references, external to the observer, physical elements which may vary widely in scale and provide spatial prominence visible from many locations where they may offer a sequence of recognizable features.

Lynch describes how different characteristics provide reinforcement of image to give different districts recognizable character 'reinforcement of clues is needed to produce a strong image. All too often, there are a few distinctive signs, but not enough for a full thematic unit. Then the region may be recognizable to someone familiar with the city, but it lacks any visual strength or impact' (Lynch, 1990).

BOX 1.1 LYNCH'S ORGANISATIONAL ELEMENTS INVOLVED IN READING THE IMAGE OF THE CITY (LYNCH, 1990)

1 Paths. Paths are the channels along which the observer customarily, occasionally, or potentially moves. They may be streets, walkways, transit lines, canals, railroads. For many people, these are the predominant elements in their image. People observe the city while moving through it, and along these paths the other environmental elements are arranged and related.
2 Edges. Edges are the linear elements not used or considered as paths by the observer. They are the boundaries between two phases, linear breaks in continuity: shores, railroad cuts, edges of development, walls. They are lateral references rather than coordinate axes. Such edges may be barriers, more or less penetrable, which close one region off from another; or they may be seams, lines along which two regions are related and joined together. These edge elements, although probably not as dominant as paths, are for many people important organizing features, particularly in the role of holding together generalized areas, as in the outline of a city by water or wall.

3 Districts. Districts are the medium to large sections of the city, conceived of as having two-dimensional extent, which the observer mentally enters inside of, and which are recognizable as having some common, identifying character. Always identifiable from the inside, they are also used for exterior reference if visible from the outside. Most people structure their city to some extent in this way, with individual differences as to whether paths or districts are the dominant elements. It seems to depend not only upon the individual but also upon the given city.

4 Nodes. Nodes are points, the strategic spots in a city into which an observer can enter, and which are the intensive foci to and from which he is traveling. They may be primarily junctions, places of a break in transportation, a crossing or convergence of paths, moments of shift from one structure to another. Or the nodes may be simply concentrations, which gain their importance from being the condensation of some use or physical character, as a street-corner hangout or an enclosed square. Some of these concentration nodes are the focus and epitome of a district, over which their influence radiates and of which they stand as a symbol. They may be called cores. Many nodes, of course, partake of the nature of both junctions and concentrations. The concept of node is related to the concept of path, since junctions are typically the convergence of paths, events on the journey. It is similarly related to the concept of district, since cores are typically the intensive foci of districts, their polarizing center. In any event, some nodal points are to be found in almost every image, and in certain cases they may be the dominant feature.

5 Landmarks. Landmarks are another type of point-reference, but in this case the observer does not enter within them, they are external. They are usually a rather simply defined physical object: building, sign, store, or mountain. Their use involves the singling out of one element from a host of possibilities. Some landmarks are distant ones, typically seen from many angles and distances, over the tops of smaller elements, and used as radial references. They may be within the city or at such a distance that for all practical purposes they symbolize a constant direction. Such are isolated towers, golden domes, great hills. Even a mobile point, like the sun, whose motion is sufficiently slow and regular, may be employed. Other landmarks are primarily local, being visible only in restricted localities and from certain approaches. These are the innumerable signs, store fronts, trees, doorknobs, and other urban detail, which fill in the image of most observers. They are frequently used clues of identity and even of structure, and seem to be increasingly relied upon as a journey becomes more and more familiar.

(Extracted from Lynch, *The Image of the City*, 1990).

Lynch emphasises that 'These elements are simply the raw material of the environmental image at the city scale. They must be patterned together to provide a satisfying form' (Lynch, 1990). In other words, it appears that urban dwellers make sense of their lived-in environment in terms of mapping the cityscape regarding its legibility derived from the dominant urban forms and characteristics of the locale. The cityscape and its constituent forms are 'read' by the individual in the context of personal experience. Strong physical forms and boundaries promote clear reference points, promoting security and identity, and allowing easy navigation and sense of identity. Lynch thus emphasises legibility and that the city is read by experience, i.e. knowledge about the city is gained by experience and exploration, focusing on what it means to the individual.

Jan Gehl – Life Between Buildings

Like Lynch, Gehl considers the urban experience in terms of urban design, but from a social perspective focusing on the public realm, exploring what he terms the 'Life Between Buildings'. In particular, he addresses the public realm by examining the opportunities for social interaction and related communication. Gehl emphasises the importance of the different senses and the importance of walking in terms of engaging with the environment, linked to the ways that public space is organised and offered to the urban user, adopting the adage 'Life takes place on foot'.

Gehl explores ideas concerning sensory engagement in relation to spatial parameters, and thus how different sized spaces relate to the human ability to communicate. Advancing understanding of the public realm is by understanding how it contains opportunities for social engagement, which in turn is dictated by properties concerning spatial dimensions and organisation, but also the interface between public and private realms (Figure 1.2). In this respect he observes some spaces to be relatively sterile, others lively, while access characteristics may inhibit or promote social interaction. Gehl's aim is to improve the quality of the environment and thus the liveability of cities by removing barriers to social engagement, optimising how the public realm provides a positive experience which underpins quality urban living.

Gehl emphasises that space characteristics must resonate with access characteristics and must facilitate interaction by the way that they are organised, avoiding characteristics that may be confusing, e.g. competing uses or confusing information. In particular, he discusses how we are influenced by the visual field in front of us, what we see at ground level, the ground floor of buildings, including how the interface between ground level structures and the viewer forms an active interface or 'active frontage'. Thus, from Gehl, it is possible to consider from the perspective of the user, that public/social space can be more properly defined by what we can see and observe rather than just by ownership or designation. The detail of how places will perform will vary and can be analysed in terms of its potential to contain different social activities, but is specifically dependent on how spaces are perceived according to the senses (Figure 1.3), how the different senses are engaged and that slowing down the speed of walking promotes deeper sensory engagement and social interaction.

FIGURE 1.2 Life between buildings: Kings Cross, London. Designed city space where the scale relates to different aspects of human movement and sensory perception which facilitates social interaction, plus the draw of biophilic components.

As Gehl (1987) states,

> Events to be perceived must therefore take place in front of the viewer and on approximately the same level. . . . As an integral part of walking this active frontage is continuously monitored by the individual.
>
> Only 'on foot' does a situation function as a meaningful opportunity for contact and information in which the individual is at ease and able to take time to experience, pause, or become involved.

Lynch and Gehl present complementary ideas regarding perception of the urban landscape, which deal with both built forms and the spaces between based upon the experience of walking and the positive, or indeed, negative flows that are engendered as one travels through the urban environment. Gehl recognises the crucial aspects of space characteristics which need to reference human scale in promoting social engagement and which are influenced by space qualities, size and boundary characteristics. Lynch describes the characteristics which define a city as legible and thus reinforcing identity. Both consider urban forms in respect of the everyday experience, and both identify that cities rarely show optimum characteristics.

FIGURE 1.3 The rhythm of walking. The environment is explored visually and then experienced by walking to realise sensory affordances. Notice the benches, which provide places of prospect and refuge.

BOX 1.2 HUMAN MOVEMENT AND THE SENSES

- Human movement is by nature limited to predominantly horizontal motion at a speed of walking. The senses are essentially frontally oriented, and one of the best developed and most useful senses, the sense of sight, is distinctly horizontal. The horizontal visual field is considerably wider than the vertical. If one looks straight ahead, it is possible to glimpse what is going on to both sides within a horizontal circle of almost 90 degrees to each side. The downward field of vision is much narrower than the horizontal, and the upward field of vision is narrower still. A person walking down a street sees practically nothing but the ground floor of buildings, the pavement, and what is going on in the street space itself. Events to be perceived must therefore take place in front of the viewer and on approximately the same level.
- Two categories of the sensory apparatus can be defined: the distance receptors – eyes, ears, nose – and the immediate receptors – skin, membranes, muscles. These receptors have different degrees of specialisation and different functional spheres. In the present context the distance receptors are of particular importance.

- Different senses operate at different ranges. The sense of smell registers variations in odors within a very limited range. The sense of hearing has a greater functional range. Within distances of up to seven meters, the ear is quite effective. It is possible to hold conversations with relatively little difficulty up to this distance. Beyond 35 meters (100 ft.), the ability to hear others is greatly reduced.
- The sense of sight has an even wider functional area. It is possible to see the stars and often possible to clearly see airplanes that cannot be heard. In connection with experiencing other people, however, the sense of sight has, like the other senses, well-defined limitations. It is possible to perceive people at distance from ½ to 1 kilometer (1,600 to 3,200 ft.). At approximately 100 meters (325 ft.), figures that can be seen at greater distances become human individuals. At a distance of between 70 and 100 meters (250 and 325 ft.), it begins to be possible to determine with reasonable certainty a person's sex, approximate age and what that person is doing. At this distance it is often possible to recognize people one knows well on the basis of their clothing and the way they walk. Not until the distance is considerably shorter does it become possible to discern the details that permit one to perceive other people as individuals. At a distance of approximately 30 meters (100 ft.), facial features, hairstyle and age can be seen, and people met only infrequently can be recognized. When the distance is reduced to 20 to 25 meters (60 to 80 ft.), most people can perceive relatively clearly the feelings and moods of others. At this point the meeting begins to become truly interesting and relevant in a social context.
- At even shorter distances, the amount and intensity of information is increased greatly because the other senses can now begin to supplement the sense of sight. At distances of 1 to 3 meters (3 to 10 ft.), at which normal conversations usually take place, the experience involves the degree of detail generally necessary for meaningful human contact. At still shorter distances, impressions and feelings are further intensified. A very conscious use of distances is involved in nearly all contacts.
- The interplay between the intensity and distance of sensual impressions is widely used in human communication. The intense emotional contacts take place at quite close range, 0 to ½ meters (0 to 2 ft.), where all the senses can work together and where all nuances and details can be perceived clearly, while the less intense contacts take place at greater distances, ½ to 7 meters (2 to 20 ft.)
- The organs of sense are for the most part designed to perceive and process the details and impressions that are received at walking and running speed, that is, 5 to 15 kilometres per hour (3 to 9 mph). If the speed of movement is increased, the possibility of discerning details and

processing meaningful social information drops sharply. It is important that all meaningful social activities, intense experiences, conversations and caresses take place when people are standing, sitting, lying down or walking. One can catch a brief glimpse of others from a car or from a train window, but life takes place on foot.

- Correspondingly, small spaces tend to be perceived as warm and personal. The small dimensions make it possible to see and hear other people, and in small spaces, the details as well as the whole can be enjoyed. Conversely, large spaces are perceived as cold and impersonal. Buildings as well as people are 'kept at a distance'.
- The relationship between distance and intensity, closeness and warmth, in various contact situations has an important parallel in the prevalent perception of architectural dimensions. In cities and building projects of modest dimensions, narrow streets, and small spaces, the buildings, building details, and the people who move about in the spaces are experienced at close range and with considerable intensity. These cities and spaces are comparably perceived as intimate, warm and personal.
- Conversely, building projects with large spaces, wide streets and tall buildings often are felt to be cold and impersonal. In addition to the requirement that objects and events be near eye level in order to be perceived, and to the requirements imposed by the limited range of human vision, another important factor in experiencing others is that there must be a reasonable amount of time in which to see and process visual impressions.

(Summarised from Gehl, 1987)

Notes: 1. The entries for this table largely cite or paraphrase the words of Gehl; accordingly, the reader is referred to the original source. 2. While aspects of social interaction, i.e. with other humans, is the focus here, reference can also be made to how the senses engage with other aspects of urban form or associated components. 3. See also UKEssays, 2018.

Active Engagement and Being in Touch with the Places We Inhabit

It is a relatively simple process to move from Gehl and Lynch's ideas to consider more specifically the rhythms involved in different patterns of access, making sense of what we see, exploring and experiencing the flows that occur between the individual and the city environment, or, as Bannon states, 'Being in touch or in dialogue with the places we inhabit' (Bannon, 2013; Lefebvre, 1991, 2004). Bannon looks at 'an everyday aesthetic for walking' and the rhythm or 'active engagement' indicated, citing Dissanayake (1992) who (writing from an anthropological stance) argues that:

> Humans are predisposed to active engagement in making sense of individual and collective experience. Perception in this case is an active process of searching for order, pattern, form and structure in a simultaneously cognitive, emotional and operational manner.
>
> . . . In shifting through space we experience an ebb and flow of persistent personal and social psycho-geography in tune with the prized harmonious succession of little joys, mingling the qualities of how we live with what we do.

It is clear that engagement is an active process, involving a dialogue between 'place' physical characteristics and the individual, how the senses provide information and how this information is processed by the individual, including the connections that are made. They reveal that how we interact and read an environment is a complex synthesis between humans and the properties of the environment being experienced. Those properties are identified by urban designers, guided or informed by principles that are derived from observing and interviewing users and informed by an understanding of sensory engagement. If the properties of spaces do not coincide with modes of sensory engagement, or fail to register due to their weak presentation or isolation, there is little resonance between people and space or city environment. Those properties of city and place which have meaning and are interpreted as positive are identified by urban designers as specifically able to engage the senses and facilitate joyful moments of experience through the act of walking (Bannon, 2013).

Theories of Landscape Perception

One of the key theorists regarding perception of the landscape is the environmental psychologist Stephen Kaplan, working in the USA (often in partnership with Rachel Kaplan) and writing extensively on the subject. Environmental psychologists, like the Kaplans, specifically consider the aesthetic concerned with the body moving through and interacting with landscape space, how they react to the scene (Kaplan, 1979; Kaplan & Kaplan, 2003). This includes how they make sense of it, what personal judgements are involved and how legible it is, but also that the process is intuitive and unconscious, i.e. that there is an initial automatic response involving an assessment of possibilities and that this influences the decision whether it is worthwhile exploring further. Kaplan (1979) emphasises that the landscape needs to be legible so that the viewer can create a cognitive map but also offer intrigue and mystery, i.e. offer the potential for some form of reward if explored further (Figure 1.4).

In Kaplan's words, 'People's reactions are viewed in terms of what sense they are able to make of the scene and what interest they are able to find in it. . . . When people view a landscape they are making a judgment, however intuitive and unconscious this process may be. This judgment concerns the sorts of experiences they would have, the ease of locomoting, of moving, of exploring, in a world of functioning, in the environment they are viewing'.

Kaplan discusses how we engage with a landscape, initially regarding 'the two-dimensional space of the "picture plane", where the assessment is in terms of

FIGURE 1.4 Recorded flood levels of the River Severn. Plaques and stones inset to show the different record flood levels of the river and serving to connect people to the power of nature and provide moments of reflection. The highest level was in 1672, and still has not been exceeded.

coherence and complexity. In addition to this "surface" analysis, there is a rapid and unconscious assessment of what one would experience if one were to proceed "deeper" into the scene . . . how legible it is likely to be and how much additional information is likely to be provided'. (Kaplan, 1979)

BOX 1.3 KAPLAN'S FOUR INFORMATIONAL ELEMENTS

Coherence. Coherence is the 'making sense' component at this surface level of analysis. It includes those factors which make the picture plane easier to organize, to comprehend, to structure. Coherence is strengthened by anything which makes it easier to organize visual patterns into a manageable number of major objects and/or areas. These include repeated elements and smooth textures that identify a 'region' or area of the picture plane; something that draws one's attention within the scene should turn out to be an important object or a boundary between regions or some other significant property.

Complexity. Complexity is the 'involvement' component at this surface level of analysis.

Mystery. One of the most striking aspects of people's reaction to landscapes that suggests a three-dimensional interpretation is their preference for scenes where it appears as if one could see more if one were to 'walk into' the scene a ways. Mystery involves not the presence of new information, but its promise. Mystery embodies the attraction of the bend in the road, the view partially obscured by foliage, the temptation to follow the path, 'just a little farther'. There is a connection between what is seen and what is anticipated.

Legibility. A highly legible scene is one that is easy to oversee and to form a cognitive map of legibility is greater when there is considerable apparent depth and a well defined space, distinctive elements well distributed throughout the space that can serve as landmarks, including the ease with which one can perceive the space as divided up into sub-areas or regions. Like mystery, legibility entails a promise, and is concerned with interpreting the space, with finding one's way; it deals with the structuring of space. Coherence concerns the conditions for perceiving, while legibility concerns the conditions for moving within the space, although the interpretation of a scene in three dimensions is, like the analysis of the visual array or 'picture plane', an automatic and generally non-conscious process. It characteristically happens very rapidly and effortlessly.

(Compiled from Kaplan, 1979; see also Kaplan & Kaplan, 1989 and Herzog et al., 1976)

From this, one can assume that if legibility is lacking or the messages are weak, further exploration of the environment is unlikely to occur, as the indications are that there is little evidence that would be of any personal satisfaction. Conversely, landscapes that stimulate intrigue and potential call out for further exploration.

Prospect and Refuge

The theory of prospect and refuge is allied to the above discussion and concerns how we orientate ourselves in the landscape, seeking areas of safety or refuge from which we view the opportunities presented by the landscape in front of us, the prospect. The theory is advanced by Jay Appleton (Appleton, 1986) to consider how we inhabit the landscape. Appleton's ideas concern how we locate ourselves in the landscape in a position of security viewing the scene and assessing its potential. Prospect and refuge emphasises the qualities of being able to observe without being seen and is discussed from the point of an informational perspective, i.e. the ability to observe the landscape and explore its properties from a refuge point, or to see what's going on. The informational opportunities being observed in the landscape are laid out in the prospect.

In the real world of human experience, prospect and refuge are demonstrated in many associations. For example, Appleton, when explaining his concept to a student audience, suggested that we preferentially walk around the perimeter of large landscape spaces, as the perimeter suggests safety and offers a position from which it is possible to explore the prospect, the view in front of us. Similarly, Johnson (2018) explores the experience of urban spaces when the prospect or scene is viewed from the security and familiarity of a park bench, where it is possible to observe events from a favoured position but which also involves social interaction. Johnson's park bench can thus be dually considered as a social place which has the right proportions to engage the user, but also possessing the right qualities of refuge to observe the scene. Based on her observations and working with colleagues from social science and landscape design, Johnson is able to come up with guidelines for best practice which essentially apply Appleton's ideas.

Affordances and Congruence

Affordances concern what the environment has to offer, i.e. assessing its potential in terms of satisfying a need. Congruence is a related concept which compares how well current experience fits with expectations and past experiences, how well they match.

The idea of affordances was developed by Gibson (Gibson, 1966, 1979; see also Greeno, 1994). Gibson believed that we do not perceive objects in the environment in terms of just their physical characteristics but in terms of their meanings, i.e. referring to what a scene has to offer as far as the individual is concerned. The concept of affordance, originally articulated in the wider context of animal responses, is based on how a species recognises the sorts of environments it functions well in and has a preference for; it is the outcome of a process that includes perceiving things and spaces and reacting in terms of their potential usefulness or how they provide support. Thus, the landscape is explored and information gathered about the environment in terms of its possibilities and what function it might serve, for example, regarding socialisation, resting or leisure-based activities.

In this sense, environmental features are experienced as having meaning for the individual; it is possible to say that individuals have a relationship with their environment underpinned by the affordances indicated, and demonstrated by the physical properties and spatial characteristics. However, if affordances are not present or are so weak that they are not recognised, then no relationship is developed. Positive affordances are potentially beneficial to the user, while negative affordances are potentially harmful, i.e. there is a polarity involved with assessing the landscape where the viewer considers attributes as indicating positive or negative affordances. It is the form (i.e. structure) of environmental features that determines what they afford to specific users.

Congruence is a term used by Carl Rogers (see McLeod, 2014) to describe a state in which a person's ideal self and actual experience are consistent or very close. When an individual perceives this congruence, there is awareness of a fit. In terms of landscape experience, this would indicate that people seek situations which correspond to and reinforce their self-image. Put simply, we seek out landscape situations which equate to our values and identity. Rogers, in discussing the idea of congruence, believed that humans have a basic motive to fulfil one's potential through self-actualisation and thus achieve the highest possible plane of being when their ideal self equates with their actual self or self-image (McLeod, 2014). We want to experience situations that are consistent with our ideal self. Where experience is consistent or congruent with our ideals and identity, the higher our sense of self-worth (McLeod, 2014). Affordance and congruence ideas are thus evident in how we read and respond to the landscape, but also suggest that there is a fundamental driver of self-actualisation in seeking out situations and realising affordances that are congruent with expectations in order to realise one's full potential. Theory suggests that the process involves several basic operating principles:

- Making sense of what one sees, including the ability to understand what is going on and appreciate its potential, to thus be stimulated into further exploration.
- Perceiving things and spaces and reacting in terms of their potential usefulness, what they might afford and how they might provide personal support.
- Seeking out situations that are congruent with prior experience and align with or support self-image; these will be the environments that would be preferred, and interfacing with them will help confirm a sense of self-worth.
- Landscape that is too simple, not offering potential or intrigue, fails in offering affordance for involvement.

It is thus also the case that, while people often appreciate specific landscapes congruent with their expectations, the fine detail of their interpretations differ; they are interpreted according to self-identity. Where the experience has clearly been sufficiently robust to stimulate a strong response, i.e. it has been possible to read the landscape, each person evaluates the experience according to their own criteria (Coles et al., 2013). This is discussed in landscape research regarding the multiple ways that landscapes come into being. Landscapes are viewed and perceived by the senses, but interpreted by the self, where they are valued, absorbed, dissected and re-assembled by selectively combining those aspects which are congruent with self-identity (Edensor, 2010; Wylie, 2012; Ingold, 2000). Essentially, while people (automatically) assess the characteristics, in terms of cognition they are made sense of in relation to self. This means that landscapes are polysemic: they can be interpreted in different ways, and they can reflect (and indeed endorse) individual experience and identity. Indeed, Kaplan (1979) suggests that while perception is the same for all people, it is interpretations that vary, but that in reality perception and interpretation are inseparable, since perception is itself influenced by all those cultural, experiential and individual factors that are supposed to underlie interpretation.

While this may seem problematic, in practice, the key aspects are that the environment has characteristics that enable it to be read, i.e. possess strength of character for the viewer to discern and respond to its qualities, it is then possible to leave interpretation to the individual in the knowledge that a situation will accommodate many different interpretations. This is particularly so regarding those landscapes which show strong biophilic features. The biophilia hypothesis (Wilson, 1984) suggests when it comes to experiencing nature, the body naturally responds to nature's cues – it is programmed to do so, and requires no conscious effort. On the other hand, environments focused on built forms lack this natural response, and may be more difficult to interpret, more dependent on cultural factors and similarly more dependent on the strength of their form. It is thus easy to see how the presence of vegetation and other aspects related to biophilia act as a mediator in the perception and interpretation of the urban environment and how this might relieve some of the cognitive load. This may go some way to explaining the difference in response to nature and city, although the situation, no doubt, is complex. Certainly, it is clear that as an urban society, as we experience less and less routine contact with nature, its mediating effect is diminished, but also that the reverse is also true. It would also seem reasonable to suggest that biophilic components provide a guiding pathway for interpretation. Since perception of landscape is specific to the individual, then one should be open minded and perhaps unconcerned or at least not alarmed about the spectrum of interpretation. Problems only occur when one assumes that all others have the same interpretation or that one group attempts to impose their perception (and often, value system) over others.

Conclusions

This chapter has discussed a selected range of theories which concern how people interact with and perceive the environment. In presenting information, we have attempted to introduce the reader to a range of theory developed by different disciplines and spanning a range of circumstances, where there is consistency regarding walking being the natural way that the environment is experienced. The relationship between the environment and the individual user, the way it is perceived, is a synthesis involving the individual's innate biological/biophilic response, and culturally derived and personal responses; in this way, meaning is constructed.

Underpinning this is that impoverished urban environments do not provide for adequate stimulation, but environments which have strong features and offer scope for exploration do. Natural or biophilic environments do this is in a natural way, where it seems that our innate biophilic roots and sensory systems naturally coincide. The human body moves (by walking) through an environment, engaging all senses, often in an unconscious manner. There is no effort in reading the natural environment; we are naturally cued in and experience a rhythm of engagement which is satisfying and restorative.

Findings from across disciplines emphasise that it is the way that the body and mind are aligned with the environment which is central to perception and is concerned with how the senses operate and the scale of interaction. These aspects have been explored in relation to the spatial organisation of the city, the social characteristics of the public realm and the detail of the perceptual process.

The indications are that there is a constant flow of information as the senses respond to different features, and it is how this information is interpreted which gives meaning to the encounter. Findings suggest that the environment is initially explored visually, and if it shows promise, explored further involving deeper interaction and deeper sensory engagement. Theories of landscape perception indicate that the landscape is interpreted or interrogated in terms of how it might support or benefit the individual, and has been discussed in relation to self, and how landscapes reinforce self-identity involving the concepts of affordances and congruence.

On a broad scale, cities are read and understood regarding how the sequencing of features, boundaries, distinctive places, nodes and landmarks and connecting paths combine. Environmental perception is also discussed in terms of the vitality of the public realm, which emphasises that places need to reflect human scale and how the senses operate and is specifically discussed in a social setting.

The significance of biophilia hypothesis has been briefly considered in relation to the perception of the landscape, and suggests a mechanism by which biophilic components in the landscape/environment act as a guiding pathway for interpretation but that cultural and natural responses combine. It is likely that pursuing the practice of enriching the biophilic qualities of urban places will indeed make it easier to perceive the landscape and reduce the cognitive load otherwise involved. We seek experiences that accord with our innate biological roots and functioning that satisfy our needs, but which support our own identities, and these appear to be drivers of the system, involving individual cultural and personal constructs. Despite this, several authors comment on the poor-quality ability of places to stimulate any meaningful dialogue, in which case the whole system will be weak and lack any relevance.

References

Appleton, J. (1986). *The Experience of Landscape*. Chichester: John Wiley & Sons.

Bannon, F. (2013). Synchronising self and city: An everyday aesthetic for walking. In Coles, R. W., Millman, Z. (eds). *Landscape, Well-being and Environment*. Abingdon: Routledge, pp. 38–52.

Barry, V., Blythe, C. (2018). Growing pathways to well-being through community gardens and greenspace. In Coles, R., Costa, S., Watson, S. (eds). *Pathways to Well-being in Design Examples from the Arts, Humanities and the Built Environment,* Chapter 4. Abingdon: Routledge, pp. 76–96.

Calabrese, E. F., Dommert, A. (2018). Biophilia and the practice of biophilic design. In Coles, R., Costa, S., Watson, S. (eds). *Pathways to Well-being in Design Examples from the Arts, Humanities and the Built Environment.* Abingdon: Routledge, pp. 97–127.

Choudhry, K. Z., Coles, R., Qureshi, S., Ashford, R., Khan, S., Mir, R. R. (2015). A review of methodologies used in studies investigating human behaviour as determinant of outcome for exposure to naturalistic and urban environments, *Urban Forestry & Urban Greening* 14(3), 527–537. https://doi.org/10.1016/J.UFUG.2015.03.007

Coles, R. W. (2014). Environmental interaction and engagement – Supporting well-being. In Burton, E., Davies-Cooper, R., Cooper, C. (eds). *Wellbeing and Environment.* Oxford: Wiley-Blackwell, pp. 445–499

Coles, R. W., Millman, Z., Flannigan, J. (2013). Urban landscapes – Everyday environmental encounters, their meaning and importance for the individual, *Urban Ecosystems* 16(4), 819–839. https://doi.org/10.1007/S11252-013-0327-Y

Dissanayake, E. (1992). *Homo Aestheticus: Where Art Comes from and Why*. New York: The Free Press.

Edensor, T. (2010). *Geographies of Rhythm*. Farnham: Ashgate.

Gehl, J. (1987). *Life between Buildings: Using Public Space.* Translated by Jo Koch. New York: Nostrand Reinhold.

Gilbert, O. L. (2011). *The Ecology of Urban Habitats.* Netherlands: Springer.

Gibson, J. J. (1966). *The Senses Considered as Perceptual Systems.* Boston: Houghton Mifflin.

Gibson, J. J. (1979). The theory of affordances. In *The Ecological Approach to Visual Perception.* Hillsdale: Lawrence Erlbaum Associates, Inc.

Greeno, J. (1994). Gibson's Affordances, *Psychological Review* 101(2), 336–342. https://doi.org/10.1037/0033-295X.101.2.336

Hartig, T., Mang, M., Evans, G. W. (1991) Restorative effects of natural environment experiences, *Environment and Behavior* 23(1), 3–26. https://doi org/10.1177/0013916591231001.

Herzog, T. R., Kaplan, S., Kaplan, R. (1976). The prediction of preference for familiar urban places, *Environment and Behaviour* 8(4), 627–645. https://doi.org/10.1177/001391657684008

Ingold, T. (2000). *The Perception of the Environment: Essays on Livelihood, Dwelling and Skill.* London: Routledge.

Intelligent Health (n.d.). *Dr William Bird – 'Prevention is Best Medicine'.* https://intelligenthealth.co.uk/prevention-is-best-medicine

Johnson, E. (2018). *Alone together–documentary filmmaking and stories of well-being in outdoor spaces.* In Coles, R., Costa, S., Watson, S. (eds). *Pathways to Well-being in Design Examples from the Arts, Humanities and the Built Environment.* Abingdon: Routledge, pp. 203–222.

Kaplan, R. (2011). Intrinsic and aesthetic values of urban nature: A psychological perspective. In Douglas, D., Goode, D., Houk, M., Wang, R. (eds). *The Routledge Handbook of Urban Ecology*. London: Routledge, pp. 385–393.

Kaplan, S. (1995). The restorative benefits of nature: Toward an integrative framework, *Journal of Environmental Psychology* 15(3), 169–182. https://doi org/10.1016/0272-4944(95)90001-2.

Kaplan, S. (1979). Perception and landscape: Conceptions and misconceptions. In Elsner, G. H., Richard, C. (eds). *Smardon, Technical coord. Proceedings of Our National Landscape: A Conference on Applied Techniques for Analysis and Management of the Visual Resource* ([Incline Village, Nev., April 23–25, 1979]. Gen. Tech. Rep. PSW-GTR-35). Berkeley: Pacific Southwest Forest and Range Exp. Stn., Forest Service, U.S. Department of Agriculture, pp. 241–248.

Kaplan, S., Kaplan, R. (1989). *The Experience of Nature: A Psychological Perspective*. New York: Cambridge University Press.

Kaplan, S., Kaplan, R. (2003). Health, supportive environments, and the reasonable person model, *American Journal of Public Health* 93(9), 1484–1489. https://doi.org/10.2105/AJPH.93.9.1484

Lefebvre, H. (1991). *The Production of Space*. Translated by Donald Nicholson-Smith. Oxford: Blackwell.

Lefebvre, H. (2004). *Rhythmanalysis: Space, Time and Everyday Life*. Translated by Stuart Eldrin and Gerald Moore. London: Continuum.

Louv, R. (2008). *Last Child in the Woods: Saving Our Children from Nature-deficit Disorder*. New York: Algonquin Books.

Lynch, K. (1990). *The Image of the City*. Cambridge: The MIT Press.

McLeod, S. A. (2014). Carl rogers, *Simply Psychology*. www.simplypsychology.org/carl-rogers.html

O'Neill, M., Hubbard, P. (2010). Walking, sensing, belonging: Ethno-mimesis as performative praxis, *Visual Studies* 25(1), 46–58. https://doi.org/10.1080/14725861003606878

Pink, S., Hubbard, P., O'Neill, M., Radley, A. (2010). Walking across disciplines: From ethnography to arts practice, *Visual Studies* 25(1), 1–7. https://doi org/10.1080/14725861003606670.

UK Essays. (2018). *Role of Human Senses in the Spatial Experience of the City*. www.ukessays.com/essays/psychology/role-of-human-senses-in-the-spatial-experience-of-the-city-7613.php?vref=1

Von Linden, E., Lymeus, F., Hartig, T. (2022). The restorative environment and salutogenesis: Complementary concepts revisited. In Mittelmark et al. (eds). *Of Salutogenesis*, 2nd edition. New York: Publisher Springer Nature, pp. 371–385.

Wilson, E. O. (1984). *Biophilia: The Human Bond with Other Species*. Cambridge: Harvard University Press.

Wylie, J. (2012). Landscape and phenomenology. In Howard, P., Thompson, I. (eds). *The Routledge Companion to Landscape Studies*. London: Routledge, pp. 127–138.

2

EXPLORING BIOPHILIA

Introduction

This chapter explores some of the different facets surrounding biophilia, in particular with respect to real-world observations and how biophilic encounters are integrated into the everyday lives of people. It does so by presenting a series of studies which consider the connections and flows that are implicated in biophilic situations and the context in which they operate, hence to address questions concerning what processes are occurring and what circumstances promote, or indeed limit, the outcome of a positive biophilic induced experience. As such, the studies take particular account of the concepts of affordances and congruence, especially in relation to how they demonstrate social and personal connections, lived through and supported by positive environmental engagement.

Four studies are presented to construct ideas, followed by a brief exploration of biophilic design and biophilic cities. These consider the complex surrounding biophilic environments and the potential positive outcomes that can be derived from different encounters addressing the reported restorative aspects of nature. Accordingly, discussion concerns the 'operational conditions' surrounding biophilia in terms of human responses, how the impacts are expressed through the experiences of individuals who simply appreciate them for what they are and how they enrich their lives. The biophilic environment and related nature components can, of course, be experienced in a multitude of ways, but, as we will see, it involves a process of 'accessing' the experience, the ways that connections are made, and made sense of, within a personal vision involving highly personal perspectives. This includes realising affordances and how encounters are set within the social circumstances of individuals, but also how the experiences expressed are articulated in the context of self-identity and social circumstances, to the extent that it is easy to see how they are part of lifestyle choices. Later, examples consider more explicitly

DOI: 10.4324/9781003099758-3

urban design contexts which include the approach of biophilic design (Kellert & Calabrese, 2015), as well as how, internationally, cities are taking up the challenge of incorporating biophilic thinking as a key aspect of their wider city profile and functioning (Beatley, 2010; see also biophiliccities.org).

A Starting Point – Modelling the Operational Complex

This first study attempts to illustrate the dynamics of positive environmental engagement based on a computer simulation which models the levels of well-being being experienced over time as individuals move around their housing estate (based upon an area known as Castle Vale, situated in Birmingham, UK). The model was developed to test ideas surrounding the impact of the landscape on individual well-being, how places are encountered and how this involved an experience which is essentially personally supportive and restorative. The context to the model is related to the theoretical perspectives discussed in Chapter 1, especially congruence and that people preferentially relate to landscape experiences which reinforce their self-identity, and that continuous exposure to such supports self-worth. Essentially, it looks at how an integrated approach to the environment, recognising its features, its quality, spatial organisation and management are all likely to improve quality of life.

Running the simulation, and after a period of time, different places on the estate begin to show peaks and troughs which reflect individual experience at different locations (Figure 2.1). The peaks are associated with movement to higher levels of well-being; the troughs are where no such change is experienced. The work was exploratory being based upon the conceptualisation of the area as an identifiable and bounded environment with a specific population and landscape, both of which could be described by data sets and recognised the potential influence of the landscape/environment on individuals. The simulation reflects ideas relating to the positive (well-being) outcomes associated with biophilic encounters (see Chapter 1), as well as ideas linked to positive psychology regarding the context of the encounter (Sparrow, 2018; Coles et al., 2018), including spatial characteristics that engender social interaction and how space is viewed (Gehl, 1987). The development of the model was one part of a range of linked investigations that sought to better understand how a local landscape may support the well-being of its 'catchment population' in relation to the biophilic and restorative properties of different landscape encounters, how different landscape characteristics and their biophilic credentials can be described as 'supportive environments', but also recognised that biophilic connections do not operate in isolation: they are linked to and operate within real-life contexts, including the circumstances surrounding the event and the purpose of an encounter.

The project speculated that the biophilic properties of an area acted as a 'buffer' capable of helping to maintaining the health and well-being of the population through its visual presence and physical access allowing biophilic effects to take place. By providing a landscape that was free to access, maintained in good

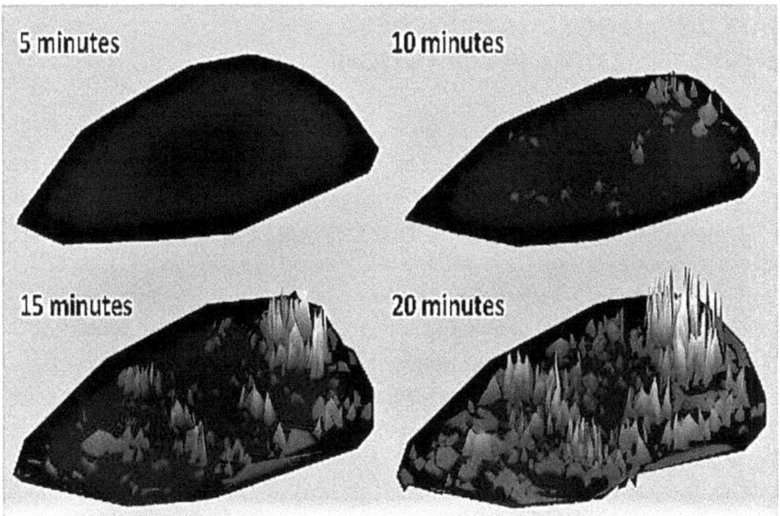

FIGURE 2.1 Modelling well-being. Levels of well-being achieved across the residential area of Castle Vale after 5, 10, 15 and 20 minutes of exposure to different environmental conditions modelled through computer simulation (see Box 2.1).

condition by the Castle Vale Community Housing Association and 'owned' by the community, the community was able to use the resource for specific relaxation and recreation purposes, but also continuously explored it when walking around the estate, or pursuing everyday social activities. The effects were reinforced by the development of the area's Central Park which hosted an annual community fair and fete. Integration of green infrastructure into the overall estate ensured consistent exposure, both casual and planned, while community feedback to the housing association helped underpin the maintenance of the estate as a proactive biophilic resource set within the wider social circumstances and walking and leisure patterns of residents.

Biophilic Encounters and Social Criteria

The second example concerns Redditch; Redditch is one of the UK's New Towns (for an example, see https://tcpa/org.uk/new-town/redditch) which followed a model of development set within a landscape complex planned at the same time as the built development takes place (Coles et al., 1995). Redditch thus saw new urban development designed as an extensive complex of housing and industry contained within a dominant landscape structure, and presented a situation where it was possible to investigate how the population responded to this comprehensive resource.

BOX 2.1 MODELLING POSITIVE ENVIRONMENTAL INTERACTION

The area being studied, Castle Vale, consisted of a discrete and identifiable residential location, being bounded by major roads and containing local facilities, including shopping and recreation. As such, it was possible to define the characteristics of the location and its population, allowing the inputting of extensive socioeconomic and landscape data.

The images are taken from different running times in the simulation, where the base levels of well-being are indicated in the first sequence and as the 'agents' roam the estate, peaks occur in certain locations.

Each agent was assigned a **well-being score** at every iteration, calculated cumulatively over walking time the score was a function of:

- **Location:** those walking in the less deprived and greener areas experience increased well-being and vice versa.
- **Group status:** as well as affecting walking speed, it was thought that those agents moving in a group experienced elevated well-being.
- **Social and repeated interaction:** this was considered important and would usually increase well-being each time agents met. Agents were given memories and could recall previous encounters to mimic events such as meeting a friend which would further inflate their score.
- **Neighbourhood:** when agents originating from the same neighbourhood met, their well-being score would experience a boost, e.g. a person seeing a familiar face, place or neighbour when out walking. When the same agents met several times, their geographical locations were pinned for the subsequent five iterations, simulating people standing together talking for five minutes.
- **Age:** in some instances, agents interacting could possibly decrease well-being score, such as a group of younger people encountering an older person. The older individual may feel threatened and experience a reduction in their score.
- The simulation takes into account a range of theoretical perspective related to congruence and the positive effects of biophilic connections and the visual plane involved in walking.

Redditch's environment is strongly defined by the inclusion of many woodland areas and parks, all easily accessible by walking and sufficiently robust to provide qualities of peace and quiet and accommodate a full range of informal leisure activities. It is possible to explore how the local population responds to the resource, since there is ample choice of locations, and thereby to decipher the biophilic connections in terms of personal reactions and experience. This involved the collection of a large number of data sets concerning how the resource was used but also what it meant to the local population in terms of quality experiences. These relate to the biophilic linkages and to the everyday activities and experiences of the population in accessing their local landscape. Notably, while individuals do express the importance of the resource in terms of nature, it is significant that use is described in terms of opportunities relating to the social circumstances of people and families, i.e. to relax, for family events, childhood play and generally chilling out. This extended across generations where the environment seemed to represent something constant and permanent in otherwise changing lives, accommodated memories and through which life events were contained or experienced to advantage. Essentially, the influence of the landscape was demonstrated as being highly therapeutic and socially connected.

Redditch was designed/planned in the 1960s as an overspill town for the nearby industrial city of Birmingham, UK. As such, it typically contains a population which can be described as culturally urban based, but that this population is exposed to an extensive complex of woodlands, plantations, parks and landscaped road verges in which dense areas of housing and industry are set. The different wards of Redditch contain a mix of residential use, industry, retail and green space, with about 15% of the total Borough area being covered by some form of public-owned woodland structure (Table 2.1). This close integration of landscape development and built development meant that the woodlands and other areas were readily and freely accessible, being under the ownership and management of the Borough Council and planted in strategic locations to separate different development areas and, in particular, along existing footpaths and pedestrian routes (as well as alongside transport routes), i.e. in the very places that people would walk or experience the landscape as a routine activity.

Redditch thus demonstrates the provision of a high-quality and robust nature-focused urban resource, and the gathered responses of the residents define the biophilic properties of this resource in terms of how they are regarded and used, including the parameters that underpin their use. Although a little dated, the Redditch study is important, as it concerns a style of landscape development that places substantial priority on the landscape as a 'common resource' with a budget to match, and which saw the establishment or retention of a large number of woodland landscapes within easy reach of the new population, meaning that there was substantial choice of what areas were available for use. In these circumstances, it was possible to establish user patterns and preferences. Importantly, these reflect the social circumstances of the population, including the frequency of use, but also the specific relevance to the individual, why they use them, i.e. the affordances realised. These are expressed by the individual and concern the characteristics of the encounter, especially the meaning concerning personal relevance and social circumstances and in relieving the pressures of everyday life.

In Redditch, the planting of woodlands was widespread throughout the Borough as a resource that could be accessed freely, i.e. when desired, without the need to make special arrangements to plan a journey or travel any distance (Coles & Bussey, 2000; Coles et al., 1995). Woodlands, in particular, are associated with strong properties which make up their biophilic profile, including the means in which the user is immersed in the environment and which offers comprehensive sensory engagement ranging from the way that light changes and the horizon is obscured, how places are revealed, to the detail of the individual trees, nature's sound, birdsong and the wind in the trees. Woodlands also reference seasonal and weather changes, changing colours and ground conditions which add to their richness in terms of fascination and exploration. Notably, woodlands are perceived as perpetual: they possess a natural dynamic which spans many (human) generations. They also operate at a scale which competes well with urban intrusion, visuals and noise, such as traffic – i.e. they enclose and insulate the user from much of the urban world, affording the experiences of getting away and fascination (Kaplan, 1995). For example, these properties and the restorative potential of the woodland experience are indicated in the practice of 'forest bathing', essentially a form of meditation which utilises the totality of immersion in the woodland environment together with exposure to a rich array of natural sensory cues, light, sound, heat of the sun in effecting the practice (see, for example, healingforest.org).

The Redditch Biophilic Experience

The relationship that individuals had with their local landscape demonstrated a simple relationship which links the ecosystem with social activities, governed by opportunities to realise affordances involving the quality of the landscape and accessibility by foot and can thus be defined by frequency of visits, preferred places to visit and the quality of the experience to the extent that it meets personal needs and expectations. If the circumstances were right, the woodlands were used as part of a daily routine and in a variety of ways as an integral part of routine lifestyle choice.

This relationship is underpinned by easy and free access, within five minutes' walking distance or less from home, plus the suitability of the environment in meeting personally defined outputs:

- **The affordances** in terms of the functional relationships, the specific aspects of an individual's lifestyle, learning to use the resource as a positive force to realise affordances, the restorative effects induced, supporting personal life strategies in ways that are congruent with their needs and expectations.
- **The quality of the experience** in terms of sensory engagement, the stimulus offered which is impacted by the woodland and associated landscape characteristics, the extent by which it 'removes' or 'isolates' the user from everyday life pressures; the presence of landscape structures which allow 'confidence of use', including familiarity.
- **Accessibility,** which is linked to the residents' ability to reach the woodland/landscape resource by walking at any desired time and without undue effort, obstacles or planning, irrespective of age or socioeconomic status and in such a way that it can be integrated with daily or frequent routines and related social activities.

At a broad level, residents expressed highly positive views regarding the landscape, reporting feelings like 'being close to nature', 'relaxed', 'happy', 'uplifted' and 'explore' – responses that suggest the ability or desire to engage with the sensory affordances offered, but also mentioned memories. 'Close to nature' perhaps needs some brief exploration in that it not only reflects the presence of nature, it also indicates the absence of urban, i.e. it is seen as a broad indicator of 'not urban', reflecting the saturation of natural experience offered and absence of urban intrusions across the range of different places.

The circumstances surrounding use are related to personal need, and affordances are identified from collected narratives which define the landscape in meeting these personal needs and the circumstances surrounding its fulfilment. There are good indications of how the experience is set within a personal context, but also that the natural environment is seen as something constant as intergenerational and timeless, i.e. spanning generations. Abbie, for example, uses the environment regarding her 'play family'; other related quotes obtained include how one individual uses the local wood to reflect and remember past relatives, while others refer to them as locations for family events, e.g. picnics and quizzes for the grandchildren. Places are appreciated for their close location and qualities, the ability to 'pop into' the wood for ten minutes or so, i.e. the immediacy of the encounter independently of their conservation status:

- 'I walk around this wood a lot, sometimes to meet my mates, but usually on my own. I cannot get a job and I am fed up. Walking around here helps me to get my head together. I do not really do anything, I just walk around. I like the peace and quiet'. (Man aged about 17 years, interviewed in Compartment 11)
- 'I like to pop into the wood for ten minutes or so when I get back from work. It helps me to unwind. It's very relaxing you know'. (Female interviewed in Pitcheroak wood)
- 'We really enjoy having a wood so close, we often go for a walk in the evening, especially in the summer . . . you wouldn't think that you were in the middle of an estate. . . . The footpaths are nice, you can even go out in the winter'. (Married couple interviewed in Compartment 11)
- 'The trees at the edge of the path are nice because when I play with my dolls I can put them against the trees so that we can have a picnic'. (Abbie, aged seven)

On site interviews, see Coles and Bussey (2000).

The Determinants of Access

The Redditch woodlands all have potential to deliver personal affordances, but there are parameters which underpin their free use and in which the realisation of personal affordances is set. Unless they are accessible and perceived as such, then the resource remains latent, never to be realised, and by preference the population requires woods to be very local, within five minutes' walking time, or 0.5km, from the place they live. At this location they can be easily accessed on foot. This relationship is quite dramatic: over 70% of all visits were to very local woodlands.

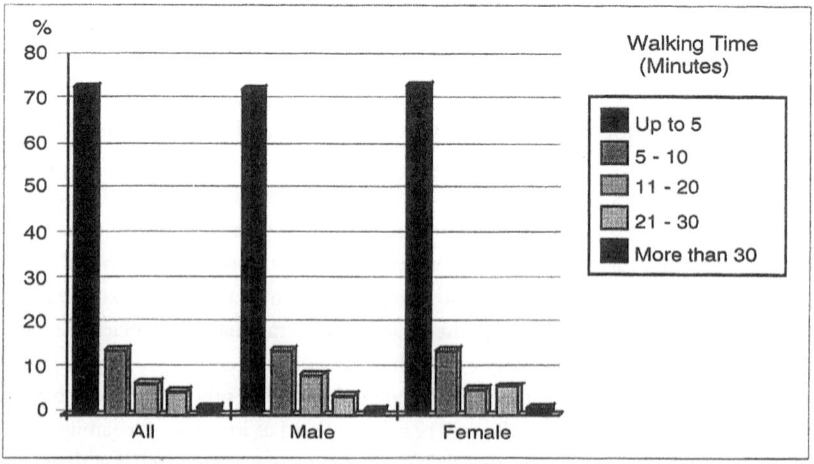

FIGURE 2.2 Green space access parameters related to walking time from the residents' homes. Bar chart showing the actual percentage of visits to green spaces in Redditch (UK) according to walking time. Most visits were local, within five minutes from the home.

Coles, R. W. & Bussey, S. C. (2000). Urban Forest Landscapes in the UK, Progressing the social agenda, Landscape and Urban Planning 52 (2–3): 181–188.

This access/distance relationship thus describes a catchment area in relation to local woodland boundaries, i.e. 0.5 km distance (Table 2.1) (see, for example, English Nature, 2003). These criteria dominated behaviour patterns, but are further influenced by the quality of the experience, determined by the ability of the landscape to isolate the user from urban intrusion involving a balance between urban noise and natural noise, to allow natural sights and sounds to dominate, offering scope for biophilic experiences to take place through multimodal sensory stimulation. Accordingly, locations with a robust structure were preferred; larger woodlands 2ha and above received more attention. A good path structure was appreciated to facilitate easy walking, while more open conditions appear to provide the right combination of ease of navigation, confident use and sensory stimulation.

That these woodland areas provided biophilic connections is to be expected from our understanding of biophilia, but understanding how their biophilic properties operate in the specifics of individual lifestyles and life stages is the crucial factor. As identified earlier, biophilic, restorative outputs are clearly demonstrated, but they could just as easily be described as social criteria, as they originally were (Coles & Caserio, 2015). They might also be described as operational or relational criteria regarding the human-environment interface reflecting the parameters of location, access, linkages to social activities and everyday activities, walking to shops, school, etc. These concern the immediacy of their use, a few moments to chill out or recover from a stressful day, and relate specifically to the lives of the individuals who experience the environment and are clearly articulated as being set within modern circumstances, especially the routine of daily living.

TABLE 2.1 Biophilic woodland provision in the different wards of Redditch that ensure equitable and free access.

Ward	*No of Woods Over 0.25Ha*	*Total Area Ha*	*Distance to Nearest Wood km*	*Distance to Nearest Wood 1-10 Ha km*	*Distance to Nearest Wood More than 10Ha km*
Batchley	8	18.90	0.5	0.5	1.0
Abbey	15	10.83	0.5	0.5	1.0
Church Hill	22	27.56	0.25	1.0	1.0
Winyates	40	48.89	0.2	0.75	0.5
Matchborough	23	43.53	0.25	0.5	1.2
Greenlands	19	35.69	0.25	0.5	0.75
Lodge Park	37	46.50	0.25	0.5	0.75
Central	15	80.51	0.2	0.2	0.2
Crabbs Cross	34	36.18	0.25	0.75	1.0
West	18	65.18	0.3	0.5	0.75
Totals	231	413.77			

Coles et al., 1995

Despite the original survey being somewhat dated, the findings still hold particular relevance in exploring biophilia and in revisiting the study to define the urban landscape in terms of human-focused biophilic properties and requirements, as opposed to relying purely on nature conservation definitions, although these are closely related, and contact with wildlife is an integral part of the biophilic experience. In Redditch, woodlands did not need to be of high nature conservation value to provide biophilic outputs, although they do need to be of an appropriate quality and nearby. The population also references these biophilic effects in acknowledging the overall visual properties of the landscape and knowing that the woodlands are there.

An Eye for the View

This investigation specifically addresses comparisons between different landscape typologies, ranging from very urban to natural, to consider how people react to and perceive different landscape situations using the approach of eye tracking. Eye tracking techniques provide a unique data stream which can be combined with interviews and commentary, during and after the eye tracking process, to understand how the viewer is interpreting the image, the two data streams being compared. Eye tracking is used in a wide range of situations to identify, for example, consumer behaviour in retail settings, but has been applied to landscape scenes to investigate further how people view and relate to the landscape using sets of different images (Choudhry, 2015). The eye tracking approach records how the eye explores an image, what point it fixes on (fixations) or how it moves from point to point exploring the image, giving valuable information concerning how the landscape is perceived. Using this technique, it is possible to compare responses to a range of different images representing views spanning natural scenes, mountains,

FIGURE 2.3A Eye tracking patterns – viewing nature. Each point represents an eye tracking 'fixation' (i.e. what is being looked at) and demonstrates how the image is viewed.

Source: For further information, see main text. Each of the associated eye tracking fixation point has additional data attached to it, allowing sophisticated interpretation. Scans patterns differ between participants, but in each participant the general pattern is consistent.

Choudhry, K., Z. (2015) A Multimodal Measurement Approach Using Narratives and Eye tracking to Investigate Visual Behaviour in Perceiving Naturalistic and Urban Environments, PhD Thesis, Birmingham City University, UK.

beaches and woodland (these can be described as biophilically saturated) and those showing strong urban influences including roads, urban scenes and industrial buildings (Figures 2.3a and 2.3b), thence to compare the results.

The results of eye tracking are fascinating since they help identify why nature scenes are preferred to urban schemes and explain the cognitive processes involved and how images of the landscape are read and interpreted.

- Emphasised is the large variety of connections and perspectives held by those viewing the same scene and which is related to the content of image as well as what people are doing in that image.
- The process unfolds from an initial response to the view demonstrated in the first few seconds (evidenced from the eye tracking data), where the image 'communicates' information to the participants; for example, the location of the first fixation indicates where the participant 'expects' to find the most interesting features.

FIGURE 2.3B Eye tracking patterns – viewing urban. Each point represents an eye tracking 'fixation' (i.e. what is being looked at) and demonstrates how the image is viewed.

Source: For further information, see main text. Each of the associated eye tracking fixation point has additional data attached to it, allowing sophisticated interpretation. Scans patterns differ between participants, but in each participant the general pattern is consistent.

Choudhry, K., Z. (2015) A Multimodal Measurement Approach Using Narratives and Eye tracking to Investigate Visual Behaviour in Perceiving Naturalistic and Urban Environments, PhD Thesis, Birmingham City University, UK.

Other connecting factors continue the process and are layered with more complexity. The perspectives, stemming from these connections, tell stories that are even more contextual in that they begin to reveal how the person understands a scene. 'Through the perspectives of like, dislike, and boring we are given a glimpse into differing world views about these landscapes, how they are seen and perceived' (Choudhry, 2015).

By comparing tracking data with the narratives accompanying the eye tracking procedure, it is possible to understand more fully how sense is made of a scene and how a scene is read:

- Participants decide within the first few seconds of exposure whether they like an image or not. Their first impression is the lasting impression and dictates whether the scene is worth investigating further.
- Initial viewing behaviour is influenced by the innate thought process and deep-rooted autonomic visual response, whereas previous experience of similar kinds of landscape influences later scanning behaviour, participants using

their past experience/memory as a template for exploring the new scene and in identifying individual landscape features.

- It is thought that humans scan environments very quickly for the reason linked to innate survival response. To survive in an ever-changing or unfamiliar environment, one needs to be quick and accurate to find out if this is a favourable environment for their survival.
- Participants 'learn' where to look over repeated exposures, but the effect of previous experience is dominant, i.e. the scenes are interpreted in the light of previous experience.

The Biophilic Scene as an 'Easy Viewing Task'

Eye tracking results can be interpreted to identify the complexity or difficulty involved in a viewing situation, and such an analysis can be applied to landscape investigation. According to Iqbal et al. (2004), the amount of time a participant's eyes are fixated on any area of interest (AOI) can provide a metric on the complexity of that area and how much mental workload it is inducing. By asking participants to view different scenes, in this case different landscapes, it is possible to observe the eye tracking patterns and how they relate to different areas of the image. In other words, this approach can be used to evaluate the cognitive complexity of a scene, where the 'patterns of eye movements can be used to highlight moments in a user's task sequence where they may have problems understanding the needed information' Iqbal et al. (2004). When this is done, there is a significant difference in eye gaze patterns when participants looked at natural compared to urban scenes; thus, it is possible to conclude that the difference in the eye pattern sequences between natural and non-natural images are linked to the complex nature of urban images. Specifically, participants show a more focused pattern in urban, and more exploratory (spread out) patterns on natural images. Since the time a participant's eyes are fixated on each area of interest relates to the complexity of that area, i.e. the time needed to make sense of it, this indicates that the viewing of urban based images was a more difficult task than the viewing of natural images. Cognitively speaking, urban images were more difficult to interpret than natural images; more mental effort was required to interpret them. Conversely, natural images required less effort and the spread-out pattern of eye movement recorded indicated that the image was easier to explore compared to urban images, more easily understood and was explored more thoroughly. The natural image sells itself more easily through easy interpretation; it facilitates entry into the landscape, inviting further exploration, and involves a more relaxing and positive task.

Perspectives, Connections and Self

Analysis of the tracking results takes account of the various data streams, the pattern of fixations and initial gaze, but can also be analysed in relation to the accompanying narrative streams (Box 2.2). These describe the diversity of

BOX 2.2 SAMPLE NARRATIVES VOICED BY PARTICIPANTS

- 'I love this image I love the flower . . . the mountain and the snow there, and the water coming down . . . and . . . it's just remind me again of my home country and at the same time I went back to Switzerland and Sweden or in Italy you can see these images but . . . this is somewhere you can breathe and you can walk freely . . . and I like that'.
- 'As a photograph I think this is a huge shame, and the road looks like a scar going through the landscape, I rather suspect if you are on the road, that it's rather stunningly beautiful because it's so dramatic and you are not aware of what the picture of the road looks like, as a photograph I think it's ugly'.
- 'I don't like this photograph so much, I think interesting architecture being dwarfed by ugly tall buildings, I can imagine there is sort of phonetic noise at ground level whether that is accurate or not, I don't know, but quite an interesting photograph, but not somewhere I particularly want to be'.
- 'As photograph, I think this is rather good it's quite dramatic with the sunset but the subject again is ugly and it's not somewhere [that] make[s] me want to be there'.

Notes: In presenting these quotes, Choudhry (2015) compares the timeline of the narrative with the eye tracking patterns to draw conclusions about how the viewing process works.
Notice the language used which identifies the properties of different scenes, and that the initial words demonstrate the immediate response which is carried through in the description as the image is explored.
All of the participants linked natural images positively to their vision and perspective, while non-natural intrusions or views negated such linkage, even though a view could be quite dramatic.

experiences and the different things that were seen as important to individuals, although their importance could be assessed as negative or positive, i.e. associated with something to avoid or something to embrace. The narratives indicated how participants were able to connect with a view, i.e. understand how it related to prior experience and then articulate the response within their own perspectives and connections. The analysis of data surrounding the viewing task, including difficulty, the initial gaze, the immediate fixation of the eye on a specific point in the view, means that the view is quickly assessed as positive or negative and processed in relation to a past remembered set of circumstances.

The Biophilic Self

These responses can be discussed in the context of the notion and reflections of 'self' developed early on in literature by geographers, which is concerned with the way that an individual views the world (see, for example, McLeod, 2014). Self is contained within an individual's cultural experience, but it also seems appropriate to recognise the 'biophilic self' (Sparrow, 2018) operating alongside the 'cultural self', which combine to act as a template for interpreting and understanding a new landscape encounter. In the light of the 'easy or complex' nature of the scanning task, 'easy' being related to biophilically saturated images, it is fairly simple to see how a natural biophilic template exists as part of self, and is used automatically in assessing the landscape scene as an 'easy task' in terms of scanning behaviour and cognitive processing. Choudhry was able to identify a range of positively linked elements in scenes by analysing all responses and which were biased towards nature references, including the categories of water, flowers, blue sky, sunlight, trees, grass and more generally relating to blue and green colours.

Choudhry stated that his work was exploratory rather than definitive, and made suggestions regarding future ideas which could provide greater clarity regarding how biophilic elements were presented/located in a scene – for example, how and where biophilic attributes are displayed, how they might influence initial gaze or facilitate further exploration of the landscape and whether the scene is worth investigating further. Although the research specifically involved images, it does follow a tradition of image use in landscape research which forms a convenient way of amassing data in a sort of laboratory setting and gives an insight into how the view is perceived, which both accords with and complements other ideas regarding landscape perception, the human condition and biophilia.

Attributes, Affordances and Everyday Biophilic Encounters

This study concerns the impact and attributes of street trees, but otherwise follows the same general trajectory of the previous studies investigating how residents attached meaning to these trees in ways that were far greater than their physical presence might suggest. Essentially, it investigates the biophilic credentials and thus significance of trees within the urban residential scene in terms of the meaning attached to them. Flannigan (see Coles et al., 2013) examined the narratives obtained from interviewing residents through 'freestyle' interviews undertaken in the home environment, typically in the living room (Kaplan, 2001), from where it was possible to observe the nearest tree during the interview, inviting participants to simply talk about their experience and thoughts. Interview responses described and defined the relationship which individuals had with these trees to effectively audit their performance and identify related affordances.

In describing the trees' attributes, the residents reference wider environmental values, especially related to nature, wildlife, the countryside and environmental

BOX 2.3 KEYWORDS EXTRACTED FROM FLANNIGAN'S INTERVIEWS WITH RESIDENTS INDICATING THE RANGE OF AFFORDANCES LINKED TO THE PRESENCE OF LOCAL STREET TREES

- Positive impact on the home environment – improving street conditions, provide shade, adding around season colour, enhance the 'feel' of the street, provide historical context, afford privacy, introduce a natural aesthetic, scenic attraction.
- Sensory engagement – the sound of nature, birdsong, rustling leaves, the sound of rain on the leaves, enhance light and shade, dancing light patterns on the wall, the smell of blossom.
- Restorative impacts – improve frame of mind, makes you feel positive, happy, calm, calming effect on the neighbourhood, reminder of happy youth, uplifting.
- Wider environmental linkages – provide oxygen, links to the countryside, help combat heat island effect, ameliorate pollution, take up carbon, the seasons.
- Nature – home for birds, attract wildlife, habitat, add nature to the street, give impression of the countryside.

Notes: These entries are a simplification of the original quotes obtained and cited by Flannigan; see Coles et al. (2013) for more detailed information.

issues, more so than had been initially expected before the surveys were analysed, i.e. the narrative that is recorded extends beyond the immediate physical presence of the trees. Clearly the experience is multisensory. It is suggested that the way that the senses are stimulated and the way that they are referenced by the individual, sets off a chain of linked associations, the street scene, and especially the trees in this case, offering sufficiently strong cues to invoke a series of associated biophilic connections. An amplification effect is thus observed in the way that the trees provide visual and sensory connection (but also historical and cultural) to more profound aspects of nature and the environment, which allows a biophilic and restorative effect to take place as identified by the interview process. In this respect, the trees are providing a pathway to more substantial environmentally focused responses.

Biophilic Design

Biophilic design advocates a series of principles which take account of the connections that are made between nature and humans, which includes patterns and features and representations of nature to advocate a suite of measures that can be applied in

FIGURE 2.4 Street trees. Street trees can present strong biophilic affordances in relation to much wider environmental issues than just their visual significance.

creating places. Specifically, these are designed to re-establish connections with the natural world but considered throughout the urban context. Biophilic design was pioneered by Stephen Kellert, and draws on the understanding of the way in which biophilia works; it is thus based on the concept specifically with the intention of creating healthy habitats for humans as biological organisms (Kellert, 2005). As a design approach, biophilic design sits relatively easily within the wider realm of urban design, seeking to provide a sensory experience which resonates with the human condition regarding our biological roots by bringing nature into the urban environment. As such, it places the integration and exploration of nature, natural systems and natural processes as a fundamental consideration in developing the urban realm and also recognises the idea of biophilic components and the organisation of space as a pathway in promoting deeper and more meaningful environmental engagement.

The Biophilic Design Framework

Biophilic design provides a set of principles (Box 2.4) which are applicable across the range of urban circumstances. Although they most clearly address design professions, they also more widely apply to such spheres as community actions and similar support initiatives, providing frameworks which can be interpreted within a range of professional, voluntary or indeed personal niches. The relationships and parameters that are indicated in biophilic connections are reflected in the ways in which individuals experience nature and natural systems, including how nature is embedded in materials and forms, as well as direct contact with natural systems, albeit often highly modified. Within this context, in seeking to restore contact with nature, it takes account of the process by which individuals encounter places, walk through them or view them, ranging from direct contact with nature, to how nature is represented or spaces organised and give an indication of where potential biophilic connections are found and thus opportunities that can be exploited. Biophilic design thus extends our understanding of urban design principles, in particular the way that nature can be accommodated and referenced in all situations, and seeks to place the individual within a set of environmental and biophysical circumstances that are capable of stimulating a biophilic reaction. It is clear that exposure to different biophilic elements offered in appropriate ways to ensure their efficacy can help underpin positivity. By recognising potential opportunities by which they can be specifically 'designed in', and addressing situations which negate their effect, biophilic design incorporates an ecology where different elements of natural biophilic orientation can be integrated into places via the design process to provide a sustained and repeated exposure to appropriate sensory cues.

Biophilic design recognises that nature can be experienced in different circumstances, the direct experience of nature and natural process but also through patterns and materials which are based upon nature and the organisation of space. These reflect that nature's processes are communicated in various ways, but also respects the ways that environments are perceived, in that those that offer intrigue and perform to a human scale are more inviting and prompt sensory responses in ways that promote further exploration and offer the promise of something worthy of attention. To this extent, it is possible to see how biophilic design is applying the power of nature in its various forms and representations as a pathway to positive engagement and, as discussed earlier, is likely to make the task of viewing the landscape an easy, relaxing activity involving an immersive and rewarding experience.

BOX 2.4 THE FIVE PRINCIPLES OF BIOPHILIC DESIGN

1 Biophilic design fosters a repeated and sustained engagement with nature.
2 Biophilic design focuses on human adaptations to the natural world that over evolutionary time have advanced people's health, fitness and well-being.
3 Biophilic design encourages an emotional attachment to particular settings and places.
4 Biophilic design promotes positive interactions between people and nature that encourage an expanded sense of responsibility and stewardship for the human and natural communities.
5 Biophilic design encourages ecologically connected, mutually reinforced and integrated design solutions.

Source: Kellert and Calabrese (2015).

As Calabrese and Dommert (2018) state:

> Upon seeing the world through a biophilic lens, the complexity and interconnectedness of our health and well-being with our physical environment become more obvious. We have unintentionally designed stress, obesity, aggression, hostility, disease, fear and environmental degradation into our lives. Quoting the late Dr. Stephen Kellert, 'We designed ourselves into this predicament and we can design ourselves out of it with the help of biophilic design'.
>
> *(in Kellert & Finnigan, 2011)*

Biophilic Cities

The logical evolution of biophilic thinking is its application at a city wide scale in which biophilic principles are understood and applied throughout the city and pervade its activities, identity and functioning, and guide the city wide agenda. The idea of biophilic cities has been pioneered by Tim Beatley (Beatley, 2010), who also set up the Biophilic Cities Network, which showcases examples of cities which have committed themselves to these ideals (biophiliccities.org).

As Beatley explains,

> A biophilic city is more than simply a biodiverse city. It is a place that that learns from nature and emulates natural systems, incorporates natural forms and images into its buildings and cityscapes and designs and plans

in conjunction with nature. A biophilic city cherishes the natural features that already exist but also works to restore and repair what has been lost or degraded.

It is our belief that every city is biophilic to some degree, and has the potential to become a more biophilic city, and that realizing that potential will require intentional changes in some municipal and urban planning and design policies and practices to produce richer, more vibrant nature-filled cities of the 21st century. The biophilic Cities Network is comprised of cities from around the globe dedicated to improving the connections between residents and urban nature.

(biophiliccities.org, Partner Cities, *accessed 02–03–2022)*

Biophilic cities have their own specific features which are recognised in terms of city profiles, but it is the political will and commitment that underpins the ideology and would suggest that biophilic solutions would be first choice regarding all aspects of operation, in the knowledge that they offer solutions that are underpinned by the power of nature and will function in a sustainable way (Box 2.5). There are also some possible comparisons with Lynch's work regarding the image of the city (presented in Chapter 1) in that biophilic cities will demonstrate their own strong identity and can be read or understood through their biophilic profile, the landscape forming a common resource which will provide particularly effective ways of engaging citizens and accommodating different interpretations and activities. For example, one of the best-developed biophilic cities, Singapore, in embracing biophilic solutions has one of the most outstanding cityscapes which involves the splendour of nature and the adoption of innovative solutions which are underpinned by biophilic ideology. The Biophilic Cities network can thus be considered as leading the way in demonstrating ideologies and structures that together provide a substantial depository of best practice.

BOX 2.5 FOUR EXAMPLES OF BIOPHILIC CITY PROFILES

CITY BIOPHILIC PROFILE

Austin Texas (USA): Nature integrated into the legal and planning process, development that prioritises vegetation, ecosystem function and the human experience, access to a daily dose of nature, connecting children with nature, Neighbourhoods Free Tree Project, 20,000 acres of parkland, 30 miles of urban trails, 38,361 acres of water quality protection, 13,610 acres of habitat preservation, City Code emphasising vegetation and ecosystem management, Imagine Austin Plan.

Birmingham (UK): Green Living Spaces Plan, linking health and green space, mapping Health Improvement Zones, 570 parks, 73,000ha of green space; extensive network of accessible canals, Multiple Challenge Ecosystem Services Maps, Natural Capital Planning Tool, Birmingham Green Commission; Nature Improvement Areas; Birmingham Open Spaces Forum; community gardens and allotments.

Singapore: Singapore city motto 'Singapore Garden City', changed to 'City Within a Garden', networks of trails and paths, integration of nature in vertical, spaces, vertical forests, roof gardens, indoor hanging gardens, increase in city green space from 36% to 47% over 21 years, National Parks Board, Garden City Fund, internationally famous botanic gardens, Center for Urban Greening and Ecology providing skills training, repository of best practice, Gardens by the Bay leisure destinations, foster pride of ownership of the City's gardens.

Vitoria-Gasteiz (Spain): High biodiversity index, over 25sq metres green space per capita, named as European Green Capital, restoration of greenbelt surrounding the city, Environment Studies Centre, urban ecological gardens promoting urban horticulture, local food production and consumption, promoting healthy eating and leisure.

Notes: The cities are defined by their biophilic attributes or credentials, plus a commitment to the natural environment through policy or action. (For full details see Biophiliccities.org.)

Conclusions

This chapter has considered a range of studies to explore how biophilic connections are realised in real-world situations in relation to the landscape experience of the individual, but also within wider city wide and design initiatives. A key observation is that biophilic connections are set within the real-world activities of ordinary people as they go about their daily routines and, in particular, relate to social circumstances. In this respect, places which demonstrate strong biophilic settings are shown to support the individual in their daily routines and can be related to sets of personal values and activities which concern self-identity and wider environmental values, and which are readily expressed by people in the form of narratives. How this occurs can be explained by reference to the theories of affordances, congruence and attention restoration theory and to the concept of self, including the biophilic self which reflects the innate link between humans and nature.

It is possible to advance the idea that the presence of nature (and nature components) in the urban scene acts as a specific positive pathway regarding the interpretation of the environment as a positive resource, wherever they occur, and that nature-based scenes are overwhelmingly linked to positive thoughts and positive actions. This is consistently demonstrated throughout different types of studies which contain strong/robust natural features even down to the scale of individual street trees; these reference the human-nature bond which can be seen across situations and circumstances. Accordingly, in experiencing the urban environment, its biophilic content is an important factor in underpinning positivity involving a complex of environmental conditions and social activities by which the urban landscape is understood and assimilated as part of personal outlook, rationalised and made sense of by cultural experience and references to such aspects as family and social circumstances.

Ideas are validated by investigation such as eye tracking, which shows the different responses to urban based and nature-based scenes regarding the effort required in their interpretation and understanding. These are based on indications that, cognitively, the viewing of natural scenes is an easy task, whereas the viewing of urban based images is a difficult task and is discussed in relation to the existence of the biophilic self. Further indication is that there is a very rapid visual appraisal of a scene which affects follow-on behaviour, but which is influenced by previous experience. In this respect, continuous exposure to nature and nature-related landscapes will stimulate the biophilic self to advantage. This situation is tested in a housing situation through computer simulation involving real-world data at a local level.

Biophilic thinking can be demonstrated at different scales of application, from local to city-wide, and is seen in the development of biophilic design and biophilic cities which specifically recognise the power of nature and utilises sets of principles that target the development of cities and places. These embrace the different facets of the complex surrounding biophilia, and seek to provide contact with nature and natural systems which are also seen in the organisation of spaces; increasingly, there are demonstration biophilic cities and innovative solutions that provide examples and exemplars of good practice.

References

Beatley, T. (2010). *Biophilic Cities: Integrating Nature into Urban Design and Planning*. Washington, DC: Island Press.

Calabrese, E., Dommert, A. (2018). Biophilia and the practice of biophilic design. In Coles, R. W., Costa, S., Watson, S. (eds). *Pathways to Well-being in Design Examples from the Arts, Humanities and Built Environment*. Abingdon: Routledge, pp. 95–127.

Choudhry, K. Z. (2015). *A Multimodal Measurement Approach Using Narratives and Eye Tracking to Investigate Visual Behaviour in Perceiving Naturalistic and Urban Environments*, (PhD Thesis). Birmingham: Birmingham City University.

Coles, R. W., Bussey, S. C. (2000). Urban forest landscapes in the UK – Progressing the social agenda, *Landscape and Urban Planning* 52(2–3), 181–188. https://doi.org/10.1016/S0169-2046(00)00132-8

Coles, R. W., Bussey, S. C., Heslegrave, W. (1995). Community forestry in an urban context. *Proceedings of the Frontdoor Forestry Conference*, Redditch, 21–22 April.

Coles, R. W., Caserio, M. (2015). *Social Criteria for the Evaluation and Development of Urban Green Spaces* (Report number D7 EVK4-CT-2000–0002). www.researchgate.net/publication/288181939_Social_Criteria_for_the_Evaluation_and_Development_of_Urban_Green_Spaces

Coles, R. W., Costa, S., Watson, S. (2018), Pathways to well-being. In Coles, R. W., Costa, S., Watson, S. (eds). *Pathways to Well-being in Design. Examples from the Arts, Humanities and Built Environment*. Abingdon: Routledge, pp. 223–238.

Coles, R., Millman, Z., Flannigan, J. (2013). Urban landscapes – Everyday environmental encounters, their meaning and importance for the individual, *Urban Ecosystems* 16(4), 819–839. https://doi.org/10.1007/S11252-013-0327-Y

English Nature. (2003). *Providing Accessible Natural Greenspace in Towns and Cities, Pub English Nature*. publications.naturalengland.org.uk/file/78003

Gehl, J. (1987). *Life between Buildings: Using Public Space*. Translated by Jo Koch. New York: Nostrand Reinhold.

Iqbal, S. T., Zheng, X. S., Bailey, B. P. (2004). Task-evoked pupillary response to mental workload. In *Human-Computer Interaction* (Proc. CHI'04). Vienna: ACM, pp. 1477–1480.

Kaplan, R. (2001). The nature of the view from home: Psychological benefits, *Environment and Behaviour* 33(4), 507–542. https://doi.org/10.1177/00139160121973115

Kaplan, S. (1995). The restorative benefits of nature: Toward an integrative framework, *Journal of Environmental Psychology* 15(3), 169–182. https://doi.org/10.1016/0272-4944(95)90001-2

Kellert, S. (2005). *Building for Life: Understanding and Designing the Human-Nature Connection*. Washington, DC: Island Press.

Kellert, S., Calabrese, E. (2015). *The Practice of Biophilic Design*. www.biophilic-design.com

Kellert, S., Finnigan, B. (2011). *Biophilic Design: The Architecture of Life*. www.filmsforaction.org/watch/biophilic-design/

McLeod, S. A. (2014). Carl rogers, *Simply Psychology*. www.simplypsychology.org/carl-rogers.html

Sparrow, J. (2018). Emotional transition and the internet world: Implications for well-being. In Coles, R. W., Costa, S., Watson, S. (eds). *Pathways to Well-being in Design Examples from the Arts, Humanities and Built Environment*. Abingdon: Routledge, pp. 140–158.

3

PATHWAYS, NATURE PLACINGS AND GREEN INFRASTRUCTURE

Introduction

The focus of this chapter is how we experience and make contact with different facets of the urban environment concerning green space and related places, especially in the context of health, biophilia, the restorative experiences and other perspectives that have already been discussed in Chapters 1 and 2. The emphasis is on people-place associations, i.e. how people are placed, or place themselves in their surroundings and the consequences of their actions in relation to their quality of life. Enquiry has thus to address both the health issues that are demonstrated in modern urban lifestyles, the tendency for less physically active routines, the stresses of urban living and the opportunities that present themselves, or need to be introduced, as well as land use priorities. As already discussed, solutions need to be embedded in the reality of urban life, the daily routine and the diversity of places, which means to fit into the patterns of how we use, move around and experience the urban environment.

At the outset, we would suggest that questions need to be asked concerning urban landscape priorities and whether existing approaches or understanding can provide the required resolution to fully address needs, and if so, how. This is a question that we will return to in later chapters, but it is helpful to introduce it now as it involves the 'where, when, why and how', exploring how nature and biophilic interactions are presented or offered (or might be offered) in the urban scene, in what circumstances and what locations.

In this regard, a useful starting position is green space and health, where understanding is relatively clear, moving progressively to consider other situations which relate to different place identities and their attributes. In doing so, we reflect on how people respond in different situations but also to consider the basic equation regarding people-place relationships, the power of nature and the human-nature bond.

DOI: 10.4324/9781003099758-4

The Position Regarding Improved Health as an Outcome

The health context of enquiry regarding urban lifestyles and access to nature and green spaces is well rehearsed. The World Health Organisation (WHO) summarise the issues: 'chronic stress and lifestyle factors, such as insufficient physical activity (Shortt et al., 2014), [and that] Urban green spaces, as part of a wider environmental context, have the potential to help address problems "upstream", in a preventative way – considered a more efficient approach than simply dealing with the downstream' consequences of ill health (Morris et al., 2006) (cited in WHO, 2016).

Accordingly, the WHO's 2016 review considers the evidence regarding 'Urban Green Spaces and Health'. The report, set within the wider perspective of public health, had an overall aim to 'help satisfy contemporary scientific standards of evidence demanded to inform policy and practice'. This is a task that appears to have been largely achieved, with the issue now turning to the development of policy and the implementation of strategies, including the setting of green space standards, involving what appears to be a shift in our understanding of the primacy of this relationship as a part of urban dynamics, including urban design (Calabrese & Dommert, 2018).

Green Space and Health, towards Standards

The health benefits associated with green space use are now well understood in that they offer qualities that promote increased physical activity, reduced exposure to air pollutants, to noise and excess heat, plus 'physiological relaxation' and 'stress alleviation'. The list continues to include improved mental health, reduced cardiovascular morbidity and mortality, a reduction in obesity and risk of type 2 diabetes and improved pregnancy outcomes (WHO, 2016). Thus, to a substantial extent, green space qualities are already defined in terms of health outcomes and thus operational objectives. However, in looking for standards and prescriptions, things are a little less clear, as the effects embrace diverse situations and circumstances; again, as the WHO state, 'the effects are heterogeneous and, therefore, they cannot be summarized in a simple fashion as a straightforward exposure-outcome relationship', i.e. a standard prescription is not yet available, although it is clear that some broad protocols can be set out and, indeed, are needed. Accordingly, the WHO concentrate on the practical realities of green space provision as an essential resource, focusing on their health-supporting properties, and aim at making a base prescription that can be adopted in practice and serve as an audit regarding the equitability and suitability of the same, that is, being consistent with health-promoting activities across different populations.

Definitions of greenspace are varied and complicate the issue, but a distinction is made between places which are primarily built forms, and urban places that are dominated by vegetation, and how they are contained within the urban scene, i.e. that they serve local populations and correspondingly should be located and be

FIGURE 3.1 Green infrastructure, St John's Churchyard. Now forming public open space. Notice the inclusion of seating and mature trees. This space allows the user to get away from the traffic as a space for both resting and moving through, depending on the task at hand, facilitating both and providing strong nature connection enhanced by the presence and architecture of the church.

integrated within residential areas forming areas for leisure. Accessibility is emphasised; for example, Natural England adopt the term 'accessible space', while the European Union specifies the context 'used predominantly for recreation'. Other definitions, e.g. adopted in the USA, provide a simpler, more basic distinction between what is green and what is not. These definitions might be considered more widely in urban situations as a more neutral starting point, without the need for qualifiers that immediately suggest that some places are excluded for consideration, and thus do not feature in the equation, which is clearly not the case. However, overall and pragmatically, they provide a workable idea of the notion of urban green from which to start discussion.

These definitions, if taken together, demonstrate green space characteristics. Key words and phrases such as – 'accessible', 'recreation', 'available for public use', 'free of charge', 'separating or surrounding areas of residential use' – indicate the ways that urban nature needs to be presented so that it can serve local populations. However, it can be argued that they are not entirely satisfactory since it is how these and other places provide or could provide a link to nature and thus support

BOX 3.1 FOUR EXAMPLES OF GREEN SPACE DEFINITIONS CONSIDERED BY THE WHO (SEE WHO, 2016, APPENDIX 1)

- The European Environment Agency 17: 'Green Space is defined as a plot of vegetated land separating or surrounding areas of intensive residential or industrial use and devoted to recreation or park uses'.
- The United States Environmental Protection Agency (US EPA) Enviro Atlas 18 (Pickard et al., 2015): 'all vegetated land, including, agriculture, lawns, forests, wetlands and gardens. Barren land and impervious surfaces such as concrete and asphalt are excluded'.
- Natural England adopts the terms 'accessible space': places that are available for the general public to use free of charge and without time restriction.
- The Urban Atlas in the European Union (Urban Atlas Code 1410): 'includes public green areas used predominantly for recreation such as gardens, zoos, parks and suburban natural areas and forests or green areas bordered by urban areas that are managed or used for recreation purposes'.

positive health-related actions, which is the issue. Green places can be highly varied in structure and size, with functions that are equally diverse, ranging from parks to urban food-growing places involving cultivation of land, through to the experience afforded by a single urban tree, or the qualities of a natural woodland encapsulated in an urban world and virtually everything in between. Similarly, within the context of 'physiological relaxation', while reference to recreation or leisure does define the broad context of use, and makes a distinction between different life activities and the various functions that different aspects of the urban environment provide, it does need a wide and flexible interpretation. Too rigid a consideration, especially where it polarises ideas, is likely to hamper fully effective green space strategies.

Accordingly, it is emphasised that green spaces should be located within easy walking distance from the home or starting point, which gives populations plenty of opportunity to access substantial green areas, and that they will be of sufficient quality to offer the deep immersive experiences required and are of a sufficient size and structure to support a range of activities related to health. Specifically, they need to respond to the needs of local populations and support the different ways in which people relate to the landscape. For example, the WHO emphasise the importance of gathering perception-based information regarding how people actually respond to and use such places as desirable (Box 3.2).

BOX 3.2 A SUMMARY OF THE PRINCIPLES OF A PRIMARY INDICATOR FOR GREEN SPACE ACCESSIBILITY BASED ON WHO GUIDELINES

- Accessibility based on residential proximity to green space sourced from defined green spaces combined with population distribution data.
- The use of appropriate green space data sources (e.g. in Europe, the European Urban Atlas).
- A green space definition of a minimum of 0.5ha as a core indicator (dependent on data resolution and availability).
- Population data at the finest spatial resolution possible.
- Proximity measured as a centroid linear distance to green space edge (e.g. edge of a park or urban woodland) involving a distance of about 300m, which approximates to five minutes' walking.
- Advantageous to carry out further surveys at a local level involving population subgroups.
- Secondary assessment of 'perception-based' indicators which address pathways to health based on the perceptions and outcomes of local populations obtained through local surveys and health-related data.

Notes: For details, see WHO (2016). *Urban Green Spaces and Health*. Copenhagen: WHO Regional Office for Europe, pp. 30–31.

Nature Placings and Biophilic Placings

Notwithstanding the critical need for guidance on and standards for green space provision, the wider issue, especially in the urban scene, is how, when and where an individual encounters nature. This acknowledges the issues of 'how we dwell with nature in the modern urban world' (Ingold, 2000), how nature is represented or presented and what effect these encounters have, including how nature can be presented in the most effective way. This essentially involves how we experience the urban landscape, how we are placed in it or how we would wish to place ourselves in it. There are, of course, boundless circumstances across the urban realm where nature is contained in some form. They concern the different circumstances and landscape typologies which offer their own dynamics of interaction, person/place relationships, across all urban situations and are essentially underpinned by a biophilic response (as explored in Chapter 2). These can be considered as situations where we are placed in nature, that is, where nature is suitably dominant and experienced directly to satisfy biophilic yearnings, but also to acknowledge that there are different ways by which nature can be embraced, all of which fall under the broad heading of 'nature placings'.

Nature placings are set within the context of biophilia and embrace green spaces. Other places, especially in the urban realm, demonstrate biophilic properties, but they would not normally be described as part of nature, but can offer qualities that are allied to the effects of nature. Rather than widen a definition of urban nature to an excessive and possibly unrealistic extent, we suggest using the additional term 'Biophilic Placings' similarly to embrace the dynamics of person-place interaction, especially since this terminology parallels the concepts of 'biophilic design' and 'biophilic cities' discussed in Chapter 2.

Another way of considering this, but from a place perspective, is how different places offer 'affordances': what they signal to the individual in terms of their potential personal use. Biophilic placings are, or can be designed to be, ubiquitous throughout the urban realm and are likely to retain their strong urban connection. Nature placings, on the other hand, are likely to be more focused regarding the intensity of immersion and involve the experience of something special, getting away, and where people can experience solitude and quietness. While they may be variable in form, they offer a needed refuge away from the everyday stress of urban life (see Figures 3.2 and 3.3). Together, biophilic placings/nature placings and affordances set a basic equation regarding person-place relationships where people make a rapid initial assessment of a landscape scene regarding the opportunity for realising personal affordances, and the extent to which they fits with expectations (see Box 3.3).

BOX 3.3 NATURE PLACINGS AND BIOPHILIC PLACINGS

Nature Placings – in relation to experiences of urban wild-spaces, the term relates to the circumstances and ways that an individual encounters and experiences nature and takes account of the effects that occur in many green spaces that are based on natural systems that in all intents can be considered as natural environments or reasonably close analogues.

Biophilic Placings – which embrace the same concept but relates to a wider consideration of how the biophilic properties of places, spaces and buildings are presented and thus experienced and the effects that occur; it takes greater account of typically designed urban forms and circumstances which are substantially removed from what is regarded as natural, also linking to the terminology of biophilic design and biophilic cities.

Green Infrastructure

Green infrastructure (GI) is a term used to describe the network of green spaces and water systems that deliver multiple environmental, economic and social values and benefits for sustainable urban development (see, for example, worldgreeninfrastructurenetwork.org, accessed 29 September 2021). The concept recognises the value that nature provides specifically to human society, and has become widely used in relation to 'ecosystem services' (Benedict & McMahon, 2006, Bennet & Chaplin-Kramer, 2016). GI is used to express the totality of spaces which specifically reference nature, as well as true nature systems and their spatial distribution where, in urban situations, they form an interconnecting network or mosaic of green, interspersed among built forms and the hard or human engineered infrastructure. It is quite usual to talk about greening solutions where natural systems and the presence of nature is actively encouraged to improve engineered systems, especially in the remit of sustainability, which essentially means moving towards replacing or softening hard solutions through the processes of nature, involving actions which can evoke synergies in terms of performance, cost savings and aesthetics, including health.

Accordingly, the GI concept and its grey/green counterpart is widely explored by a range of sectors (see, for example, the UK Green Building Council, 2015), with examples of how to integrate green infrastructure into the urban landscape. Ideas concerning GI now form a significant aspect of sustainable development practice including retrofitting, as the 'power' of natural solutions is recognised and in recognising the importance of these elements in the delivery of 'Ecosystem Services' (Bennett & Chaplin Kramer, 2016; Benedict & McMahon, 2006). Accordingly, the terminology has become mainstream in planning circles, but is readily explored by various disciplines in creating places where people want to live, and in developing sustainable solutions, e.g. regarding urban drainage, climate amelioration, pollution mitigation and other urban issues (English Nature, 2009).

> The provision of green infrastructure in and around urban areas is now widely recognised as contributing towards creating places where people want to live and work.
>
> *(publications.naturalengland.org.uk)*

Throughout the GI network, there will be plenty of situations that conform to the idea of nature placings which can be considered accordingly, and also in relation to the specific urban green space standards set by the WHO, including how they link to and can serve population centres. For example, English Nature give a Typology of Green Infrastructure which summarises some typical components. These can be regarded as different niches within the wider context of a biophilically defined urban ecosystem.

English Nature's typology covers both urban and rural areas and, in reality, would be described by the features present, say in a city. Thus, green infrastructure typologies will show the characteristics of a city or urban region and make a major contribution to the image of a city regarding identity and meaning. Several points are worth emphasising:

1 That the term covers a wide range of spatial forms, and that these all express the qualities of nature in some aspect.
2 As such, they have biophilic properties which are likely to promote human connections with nature and associated actions in ways which are positive.

BOX 3.4 ENGLISH NATURE'S TYPOLOGY OF GREEN INFRASTRUCTURE

- **Parks and gardens** – urban parks, country and regional parks, formal gardens.
- **Amenity green space** – informal recreation spaces, housing green spaces, domestic gardens, village greens, urban commons, other incidental space, green roofs.
- **Natural and semi-natural urban green spaces** – woodland and scrub, grassland (e.g. downland and meadow), heath or moor, wetlands, open and running water, wastelands and disturbed ground, bare rock habitats (e.g. cliffs and quarries).
- **Green corridors** – rivers and canals including their banks, road and rail corridors, cycling routes, pedestrian paths, and rights of way.
- **Other** – allotments, community gardens, city farms, cemeteries and churchyards.

Notes: adapted from *Green Infrastructure Guidance* (English Nature, 2009).

3 The totality of GI can be regarded as a health resource referencing the same health promoting aspects described, for example, by the WHO, which greatly extends the nature of the resource.
4 Specific reference is made to designed places as well as re-purposed transport corridors as integral components of GI.
5 Substantial reference is made to recreation opportunities.
6 The GI concept fosters the integration of natural solutions and nature places as a key part of urban form, i.e. rather than polarising what is designated as green space and what is not.
7 GI contains numerous opportunities for greenspace interaction involving nature placings, irrespective of how the places may be defined.
8 The concept is a central part of sustainable development.

Green Infrastructure and Health

Douglas et al. (2018) examined GI provision specifically as a natural health resource to develop a range of GI typologies linked to 'life stages'. These reflect different aspects of activity and related need with the aim of giving 'clear design and management frameworks'. The study was made in the context of changed lifestyles and the continued erosion of green space provision and accessibility to the same. The typologies reflect how GI is presented to local populations, with the desirable attributes linked to specific health issues and outcomes. While there are universal cross-cutting themes regarding GI attributes and provision, there are also differences in requirements which reflect engagement across the life course and specifically in relation to the implementation of urban GI strategies.

> In endeavouring to create resilient and adaptable places, the multifunctional potential of green infrastructure is increasingly recognised. The GI approach represents a holistic ecosystem approach, which includes not only the aesthetic enhancement of urban greenspaces, but also the enhancement of human health and well-being through the protection, improvement, and design of green infrastructure which is suitable and appropriate for the neighbourhood, city or region.
>
> *(Lennon & Scott, 2014, cited in Douglas et al., 2018, p. 68)*

Douglas et al. were able to construct these typologies based on needs and by specifying criteria which underpin their performance as a health resource. These identify the various subtleties needed to place the user in situations that are likely to afford the most effective health support as part of daily routine. The work effectively linked affordances to nature placings, taking into account the dynamics of urban living and life stages and how access to urban green is promoted through a natural aesthetic involving views of nature, encouraging actions and facilitating the generation of healthy behaviours as part of routine.

Five life stages were recognised – 'prenatal', 'childhood', 'adolescence', 'adulthood' and 'later life' – and observations were made regarding the different associations with positive health outcomes, and thus the required attributes of GI interventions (Douglas et al., 2018, 2017). These prescriptions take account of the needs of individuals, which can be tracked back to physical and mental health issues such as addressing inactivity and relaxation, but also suggest incorporating areas for quiet contemplation, sitting, resting and viewing nature and linked to social activities. For example, the needs of those in later life concerned addressing the possible impacts which are often associated with advanced years, including limitations on physical activity and social isolation, while for expectant mothers, exposure to greenness was linked to improved birth outcomes, with a range of interventions specified such as maximising views of green, encouraging safe and accessible walking routes and the provision of park furniture and amenities.

Nature placings are thus indicated throughout the GI network, and since this pervades an urban area, makes a major contribution to the overall greenness or naturalness of a location and is accessed on an everyday basis, visually or physically as part of walking activities. Thus, people are placed in nature as an everyday aspect of daily routines, prompting the exploration of nearby more extensive green spaces with their additional potential for health-related activities (Figure 3.2).

FIGURE 3.2 Green infrastructure in residential development. Green infrastructure developed as part of the residential landscape, concentrated on walking and cycling routes, but also to soften the impact of the road network and vehicle noise.

Alternative Pathways

How nature is experienced can be expressed in terms of different pathways: passive or immersive exposure – living or being in areas of natural greenery and experiencing the view – and active interaction, such as walking in parks, gardening, or conservation activities. In reality, these simply represent different ways of being placed or placing oneself in nature, and thus the various ways that connections are made and affordances realised. Some pathways emphasise physical exertion, (walking, climbing, digging, tending), others emphasise relaxation (resting, sitting, looking at). They relate to how people wish to engage, but multi-modal engagement is the norm in that physical exertion and mental restoration typically go hand in hand, and that a range of integrated outcomes are demonstrated.

For example, Rogerson et al. (2017) looked at the impacts of practical (physical) conservation activities in green spaces associated with volunteering work with the wildlife trusts organisation, an organisation which exists throughout the UK (similar organisations exist internationally). The report identified a combination of benefits, including the positive impacts on mental well-being and physical health, but also included enhanced pro-environment related behaviours and increased social engagement. The findings suggested that the combination of activities, involving exposure to nature and practical work in nature, and being with others with a similar outlook, were mutually supportive and also promoted common ideals concerning the stewardship of nature.

These types of initiatives are not necessarily urban based, although the participants are, but others are specific to urban areas. For example, systems such as Continuous Productive Urban Landscapes (CPULs, see Viljoen & Bohn, 2009) link the idea of urban green infrastructure with urban agriculture as a system which combines the synergies of growing food, immersion in the landscape, working with like-minded others and involving moderate levels of physical activity. Approaches like these may well define an optimum combination of physical activity and aesthetic experience desirable as a fundamental property of green spaces and which works at an appropriate level in providing connected networks of places that have the right aesthetics concerned with visual and physical activities. Such ideas present alternative urban landscape systems and suggest an alternative dynamic for the urban landscape with opportunities for both physical and restful activities integrated through appropriate landscape forms. This is a dynamic which would address both human and ecological health and urge wider innovative 'systems thinking' as the way forward (Rayner & Lang, 2012; Barry & Blythe, 2018).

Studies are thus continuously adding to our knowledge regarding how different parts of the GI system operate within an overall holistic ideal. These involve different niches of operation which address health, explore different pathways and give an indication of the diverse circumstances surrounding urban nature and greenspace. As Barry and Blythe (2018) observe,

> This is not just about the value of green spaces as an 'add on', subjective view or lifestyle 'choice', but goes to the heart of health inequalities, encompassing ways in which to support human resilience to the drivers of ill health, including tackling disparities in access to and use of green space.
>
> *(Barry & Blythe, 2018, p. 76)*

For example, the evidence for health benefits of green space related to gardening and food growing addresses a number of improved health outcomes which may be increased by the physical, emotional and social activities concerned, to the extent that Van den Berg et al. (2010) make the case for green space use to be incorporated into 'health service pathways' (Barry & Blythe, 2018; Natural England, 2010). Benefits are wide ranging, and include: increased physical activity; improved fruit and vegetable consumption and weight outcomes (Schmutz et al., 2014; Buck, 2016; Alaimo et al., 2008; Zick et al., 2013); enhanced recovery from illness; improved functioning among those with dementia; and physical and mental health benefits for the elderly (Hawkins et al., 2011; Adevi & Lieberg, 2012; Hewitt et al., 2013). Accordingly, there are a number of green health therapies that recognise this relationship, and which are forming a more significant part of health care (Box 3.5).

BOX 3.5 SOME TERMINOLOGIES ASSOCIATED WITH GREEN HEALTH THERAPIES

- 'Forest bathing' (Shinrin Yoku, Japan)
- Horticulture therapy (as a professional association in the USA)
- Care and social farming (across Italy, Norway and Belgium)
- Eco-therapy
- Green care
- The Green Gym (TCV registered term)
- Social prescribing
- Green prescribing
- Civic authority-designated 'Local Health Resources'
- Nature therapy
- Walking for health
- Green therapy
- The Natural Health Service (adopted in reference to the UK's National Health Service, see www.intelligenthealth.org)

Note: Compiled from various sources including Barry & Blythe, 2018; see also Hassink & Van Dijk, 2006; Lee & Maheswaran, 2011; Bragg & Atkins, 2016; Bragg & Leck, 2017).

For example, Barry and Blythe (2018) give examples of partnerships between health agencies and community or voluntary groups (in the UK, the National Health Service) which more formally recognise green health therapies as a legitimate and inclusive part of health care, where schemes address clinical needs, such as aiding patient rehabilitation or helping treat dementia, where ideas are becoming more firmly embedded within professional thinking and approaches. These aim to link patients with non-clinical need to wider voluntary and community support, such as joining a 'Green Gym', park walking groups or community gardening programmes (Barry & Blythe, 2018; Bragg & Leck, 2017). Schemes operate in the context of improving health outcomes as well as addressing social and community aspects. One such example discussed by Barry & Blythe concerns the example of the Queen Elizabeth Hospital in Birmingham, UK (see Box 3.6); other schemes more broadly integrate community food growing with health, and can assume sufficient significance that they receive formal recognition as 'local health resources' (for example, see Jasmine Road Gardens, available at https://jasmineroadgardens.wordpress.com/)

Chilling Out

Urban nature is more complex than just green spaces, and thus less easy to define, but this also means that in urban situations there is a much more diverse range of circumstances and places where nature is contained. These places reference nature

BOX 3.6 GROWING FOR HEALTH: COMMUNITY FOOD GROWING SPACE, THE EXAMPLE OF QUEEN ELIZABETH HOSPITAL, BIRMINGHAM, UK

- The land (over 1,600 m^2) surrounding the hospital includes wildflower and hay meadows, woodland, orchards and food growing spaces created through partnerships of residents, patients and voluntary groups, supported by corporate and public funding. Green spaces and trees provide peaceful, therapeutic spaces for staff and visitors away from the stresses of the busy hospital. Links are developing between the green spaces and the hospital's health professionals, giving them support to make use of the grounds as a resource for rehabilitation and dementia programmes. The contrast between the hospital building and its linked green space presents a more holistic view of health and well-being, demonstrating preventative approaches as well as acute care.
- Creating the community growing food space within the hospital grounds was made possible through the input of The Trust for Conservations Volunteers' (TCV) 'Health for Life' in the community programme funded by Mondelēz International, one of 15 Health for Life garden and growing 'hubs' across south Birmingham, supporting the programme's aim to increase physical activity, improve understanding of healthy diets and provide opportunities for people to grow their own food, responding to health issues in the city, including obesity, diet-related illness and physical inactivity.
- Working with residents, hospital staff, corporate sponsors, volunteers and schools, TCV created a garden 'hub' at the hospital, growing fresh vegetables in raised beds. Weekly 'Green Gym'® sessions, run by TCV staff, combine physical activity and outdoor work, gardening, cooking and eating. Volunteers from all ages and backgrounds, including those with mental health issues, poor health and long-term unemployed, use the gardens. Activities embrace the wider site with maintenance of the woodland, orchard and hay meadows with plans to extend food growing areas in order to supply the hospital canteen, modelling sustainable production.*
- Participants engaging in the Health for Life programme complete 'Health Questionnaires' covering physical activity levels, healthy eating and gardening skills. Baseline comparisons of the programme's effect indicated that 84% of participants had become physically more active, numbers increasing their exercise regimes almost doubled, 91% grew their own food as a result of involvement, 68% consumed more fruit and vegetables, 87% of participants increased social activity (made new friends) and 86% spent more time outdoors.

Notes: Information summarised from Barry and Blythe (2018), for full information refer to the original source. *Notice the combination of activities, food growing and voluntary conservation work plus the context of food preparation and cooking.

in that they have biophilic properties, but may be described in terms of urban design (and increasingly by biophilic design), typically falling within the public realm where they are understood in relation to the work of urban theorists. They assume particular importance since they permeate the urban fabric close to places where people live and work, although they might only be marginally described as green space. Exploring such spaces through multidisciplinary perspectives/collaboration – for example, combining the social sciences, landscape architecture, the arts, and urban design – reveals their value to local people, places to meet, social places, but also the environmental determinants at work that underpin the properties of place including their biophilic properties, i.e. how they form biophilic placings of substantial importance locally.

As an example, we can return to the 'Bench Project', briefly mentioned in Chapter 1 as an example of a situation involving prospect and refuge. In this example, Johnson (2018) essentially examined the affordances offered by the humble bench, and the project consortium was able to develop best practice guidelines regarding how benches should be provided in the urban environment and which recognised that the siting or placing of benches takes account of the wider biophysical characteristics of spaces, especially regarding the view, but connected community, provided a social space and could also be identified as a local health asset. As Johnson states:

> Sitting on benches supports healthy everyday routines by enabling people to spend longer outside. These opportunities to rest can be restorative for mental health and support local walking when personal mobility is limited. Benches function as a social resource, and are flexible and affordable places to spend time. This is appreciated by many, and especially vital for many people who are largely marginalised from other collective environments such as work, cafes, and educational or leisure facilities. They are contrasted positively with crowded, lonely or boring home situations.
>
> *(Johnson, 2018, p. 213)*

Johnson (2018) captured 'everyday stories of well-being in outdoor spaces' through the use of filming, film-making and interview, where the recorded oral testimonies describe both affordances and the restorative experience that took place – to identify the thoughts of frequent users 'to articulate feeling and being'. The bench was essentially a good place to be and offered isolation from the stressful world, and users were clearly moved to an alternative restful dimension. Participants' quotes identify aspects of the experience; on the other hand, site descriptors, that is, the biophysical components and spatial arrangement of the landscape, immediately show that biophilic attributes are inherent in the places examined, where the bench provides the context for their realisation the invitation to sit, relax and chill.

> It feels peaceful here – I forget about issues at home – it's a good place to like – you just come here to chill – it makes me feel happy – she loves because of the sense issue – she takes her shoes off, she likes to be on the grass.
>
> *(participant in the Bench Project; see Johnson (2018) for full details)*

Analysis of the bench project demonstrates a collection of biophilic associations, that constitute the placing, giving the necessary cues which assist, if not fully determine, how the prospect/scene is read and embraced, where they are sufficiently strong to stimulate a response. In many respects, these may be regarded as a familiar cameo biophilic situations, sitting under a tree, listening to the birds, looking at the water, feeding the ducks, watching the sunset. Once appreciated for what they are, it is obvious that such situations, or opportunities for such placings, readily present themselves throughout the urban scene but also form the special moments when immersed in deeper green experiences involving more extensive sites. Prospect and refuge situations are indicated in many viewing situations – the view from the window, looking out on nature – while biophilic encounters typically feature when sitting and resting or moving through the urban outdoors, where they offer a preferred aesthetic experience and associated opportunity to 'sit in' or 'walk through', 'take in the view' or 'soak up' the atmosphere.

Biophilic placings and nature placings are inherent in biophilic design (Calabrese & Dommert, 2018), and studies of places like these suggest that the biophilic self is highly sensitive and responsive to natural cues when they are set within an urban world, especially if they are carefully applied.

An Adequate Living Environment

An adequate living environment balances sensory inputs and provides a mix of responses that are congenial and consistent with people's culturally and evolutionary conditioned needs to re-affirm identity (Astbury, 2013; Ingold, 2014), including our biophilic identities. We fulfil this ambition in responding to landscapes that are congenially and culturally satisfying and which are congruent with our expectations. We evaluate places according to acquired information, and by preference will dwell in these places that match our value systems (and which signal the right affordances). In respect of an adequate living environment, we still have a yearning to dwell in nature. This does not imply any penalty since, as previously stated, we are highly sensitive to situations where nature is present and discernible and we can dwell within its parameters, both in natural sites, but also in places which demonstrate biophilic properties if they are appropriately offered through nature or biophilic placings. Essentially, the places where we spend our time, inside or outside, should be consistent with our needs, including our evolutionally determined needs, as indicated in the biophilia hypothesis.

While natural settings are often the preferred destinations for extended restoration programmes, idyllic places for 'getting away or being away' (Kaplan, 2001; Sawang & Guaralda, 2018), the sense of being away, and the associated restorative experience, does not require the setting to be distant (indeed, opportunities should be available locally, including indoor work environments (Jahncke et al., 2011; Sawang & Guaralda, 2018). Sitzer (2013), for example, looking at office workplaces, developed the idea of a 'restoration room', described as 'a simple room with

FIGURE 3.3 Biophilic placing as a chill-out space. An example of a 'biophilic placing' in a busy high-street, offering refuge and a restful moment among the blossom and light shade of an avenue of ornamental pear trees and allowing the pedestrian to get away from the immediate stress of things. This might be regarded as a 'chill-out space'.

a nature scene, sounds of nature, subdued lighting and a comfortable setting can be effective in reducing subjective work stress' (see Sawang & Guaralda, 2018); ten minutes' exposure was found to be effective.

Sawang and Guaralda (2018) further explored the concept and examined how the sensory properties of nature experience can be replicated by technology in

FIGURE 3.4 Accessible wild space: experiencing nature. Native woodland located adjacent to residential development, highly accessible and of vital importance, offering opportunity for deep sensory engagement. This is a large woodland and is supplemented by new plantations nearby to provide even greater opportunities for recreation and relaxation. As a wild place, it allows people to immerse themselves in the unique qualities of nature.

the form of an isolation pod that combines an appropriate mix of biophilic cues, including images, sounds, smells (for example the sound and smell of rain), humidity and temperature. The concept was that the pod could be installed in any location and takes account of the way that the senses provide 'stimuli at affective and cognitive levels. At cognitive level, individuals interpret a situation/object based on its semantic and aesthetic qualities' (Hassenzahl, 2004, cited in Sawang & Guaralda, 2018).

The indications are that by providing an appropriate mix of biophilic cues, a restorative effect can be achieved both indoors and outdoors, and while it is not clear to what extent prior experience of nature is required for the experience to maximise the effect, it seems that the technology can provide very specific fine-tuned prescriptions, although, of course, nature does so without the need for technological input. The idea of a restoration pod does mean that all the sensory inputs can be optimised in terms of quality, offering maximum potential for restoration and possibly altering the perception of time/timelessness which is also recognised as a sensory modality, and indicated in the daydream state associated with mental restoration.

The Context of Sensory Engagement and Memory

Different people identify different affordances from the same place, and it is clear that while the same sensory cues are involved, they are interpreted in the context of past encounters which provide a reference point. It is also evident that biophilic properties are demonstrated throughout the wider built environment and that different elements impact on sensory engagement. As discussed in Chapter 1, this is explored by urban theorists, especially Gehl and Lynch, regarding ideas around the image of the city and how we move and socialise in the public realm. Emphasised by these and others is that people respond in their own individual ways, although there are broad common responses. The complex of urban places, including natural land-forms and designed places, thus offer different subtleties for engagement, for realising different personally identified affordances, as well as different modes of experience involving different sensations.

These experiences affect the human psyche through direct sensory impacts and are influenced by memories of other places and innate responses linked to biophilia (Coles et al., 2013). Qualitative studies on landscape perception (as advocated by the WHO) are thus informative, where they reveal an 'attachment to place' perceived as dialogue or sensation (Gray, 2003; Massey, 1995; Ingold, 2014). For example, Millman examined how people perceived and reacted to Birmingham's central canal area (Millman, 2013, Coles et al., 2013) which features as a strong part of the city's identity. Using the concept of sending a postcard which featured a choice of images to a friend, she asked participants to select the images that they most preferred, together with a short explanation written on the reverse regarding the rationale for selection. The study provoked strong responses and it was clear that favoured images were selected on the basis of personal affordances, what they meant and how they related to their own identity, but also on the basis that they contained what might be considered iconic landscape components, including prominent trees, expanses of water, views, prospect and refuge situations, significant buildings that contained meaning, bridges and heritage. The most favoured images were also the most complex, containing a wide array of sensory cues, and accordingly elicited the most diverse responses, in that they had the capacity to engage the observer. The presence of large dominant trees appeared particularly significant, while water featured as something natural or spiritual and was a pervading biophilic aspect of the different images.

Further investigation involving walking and talking identified the importance of being in the landscape when attempting to uncover its meaning to the individual, due to the opportunity for multi-modal sensory or total body experience (Pink et al., 2010), but also demonstrated the process of memory recall and, although certain images were favoured, the reasons behind their choice were diverse and indicated an association with other landscapes and events. How people reacted to these places involved a cyclical loop of sensory cues and memory recall, including innate biophilic memory, where memories of other places were afforded by these cues. To the outsider these could seem somewhat bizarre, for

FIGURE 3.5 Birmingham City Centre Canal at Brindley Place: the canal network provides a refuge away from much of the hustle and bustle of city activities, and does appear to function as a biophilic space, largely because of the dominance of water, the absence of traffic and the variety of sensory engagement offered, but how it is interpreted is highly variable.

example, the feeling of cobblestones underfoot led one participant to recall the streets of Turkey which were similarly paved, while another was transported to a memory of the seaside near his home in Italy, by the sounds of water moving through a canal lock.

Studies such as these emphasise the importance of memory, including biophilic memory, in how people perceive places, and in particular how different facets of the built environment can cue different associations which can move the observer to a special place away from the previous reality of where they are or just come from. In the case of Birmingham's central canal-scape, this was assisted by the fact that there are no cars, the sound of traffic is absent or subdued, the presence of water is dominant and reinforced in the views along the canal. Opportunities to alter viewing perspectives, e.g. being on the water, crossing via bridges, wharves or locks, offer different modes of viewing, orientation, isolation and physicality, places for sitting and exploration by walking, while the scale of elements is consistent with human movement.

This is of course the stuff that urban design is made of, but in the context of this exploration, we would emphasise the importance of natural cues and that these excite the biophilic self and promote further exploration. The indications are that where places have an appropriate degree of biophilic saturation, the response is that 'these are good places to be' and invites further exploration which of course conforms to the ways that we perceive and react to the landscape.

Conclusions

In this chapter, we have explored environmental and biophilic interactions in terms of person-place relationships using selected examples to explain the idea of biophilic placings and nature placings. These terms simply acknowledge how people orientate themselves in the environment and experience nature and natural processes, including the qualities of other places where biophilic components are situated, in ways that they can be fully appreciated by the senses. Much of our understanding comes from studying urban green spaces which retain a structure that closely resemble natural systems. These 'nature places' offer restorative experiences and a 'sense of being away'.

Examples have been presented to try to widen an understanding of nature and biophilic placings in public places which moves discussion into the realm of urban design. It is also clear that landscapes have multiple meanings and that the related affordances are specific to individuals, although there are broad commonalities. However, we suggest that at the heart of many of these experiences is a biophilic response which is associated with restorative experiences and moving to a restful state physiologically and mentally, as indicated in the biophilia hypothesis. It appears that this response is sought out in places associated with recreation, but is also inherent in a wide range of outside places that are associated with social encounters or a moment of chilling out. As investigation moves from natural green space to other urban places, the situation becomes less easy to explain, but still points to biophilic connections as being a critical factor.

Urban biophilic places demonstrate properties where the biophilic components are sufficiently robust to provide a sense of being away, even for short periods of time, and are able to isolate the individual from the stresses of life. Perception surveys show how they are interpreted as positive therapeutic and social places that offer interactive experiences. It is clear that there are plenty of opportunities for making biophilic connections throughout the urban realm, and this can be considered in relation to how we need to dwell in nature in the urban world, especially from the point of view of health.

If the different situations for experiencing biophilic connections are considered as niches of operation, as part of a wider system, then there is an ecology of biophilia expressed in relation to human need and reactions. Biophilic and nature places thus present different niches of interaction which together constitute the totality of the biophilic infrastructure. Some niches facilitate chilling out, others clearly promote more intense physical activity, but discussion emphasises that biophilic connections should be and can be embedded as a fundamental property throughout the urban realm and that these properties are now becoming more routinely integrated into health provision.

References

Adevi, A. A., Lieberg, M. (2012). Stress rehabilitation through garden therapy: A caregiver perspective on factors considered most essential to the recovery process, *Urban Forestry & Urban Greening* 11(1), 51–58. https://doi.org/10.1016/J.UFUG.2011.09.007

Alaimo, K., Packnett, E., Miles, R. A., Kruger, D. J. (2008). Fruit and vegetable intake among urban community gardeners, *Journal of Nutrition Education and Behavior* 40(2), 94–101. https://doi.org/10.1016/J.JNEB.2006.12.003

Astbury, J. (2013). Interactive urban landscapes for well-being and sustainability. In Coles, R. W., Millman, Z. K. (eds). *Landscape, Well-Being and Environment*. Abingdon: Routledge, pp. 72–86.

Barry, V., Blythe, C. (2018). Growing pathways to well-being through community gardens and greenspace. In Coles, R. W., Costa, S., Watson, S. (eds). *Pathways to Well-being in Design Examples from the Arts, Humanities and the Built Environment*. Abingdon: Routledge, pp. 76–96.

Benedict, M. A., McMahon, E. T. (2006). *Green Infrastructure*. Washington, DC: Island Press.

Bennett, E. M., Chaplin-Kramer, R. (2016). Science for the sustainable use of ecosystem services, *F1000 Research* 5, 2622. https://doi.org/10.12688/F1000RESEARCH.9470.1

Bragg, R., Atkins, G. (2016). *A Review of Nature-based Interventions for Mental Health Care* (Natural England Commissioned Report 2014). London: Natural England. http://publications.naturalengland.org.uk/publication/4513819616346112

Bragg, R., Leck, C. (2017). *Good Practice in Social Prescribing for Mental Health: The Role of Nature-based Interventions* (Natural England Commissioned Reports, Number 228). New York: Natural England Publications. http://publications.naturalengland.org.uk/publication/5134438692814848

Buck, D. (2016). *Gardens and Health: Implications for Policy and Practice, Report Commissioned by the National Gardens Scheme*. London: The King's Fund. www.kingsfund.org.uk/publications/gardens-and-health

Calabrese, E. F., Dommert, A. (2018). Biophilia and the practice of biophilic design. In Coles, R., Costa, S., Watson, S. (eds). *Pathways to Well-being in Design. Examples from the Arts, Humanities and the Built Environment*. Abingdon: Routledge, pp. 97–127.

Coles, R., Millman, Z., Flannigan, J. (2013). Urban landscapes – Everyday environmental encounters, their meaning and importance for the individual, *Urban Ecosystems 2013* 16(4), 819–839. https://doi.org/10.1007/S11252-013-0327-Y

Douglas, O., Lennon, M., Scott, M. (2017). Green space benefits for health and well-being: A life-course approach for urban planning, design and management, *Cities* 66, 53–62. https://doi.org/10.1016/J.CITIES.2017.03.011

Douglas, O., Scott, M., Lennon, M. (2018). Green space and urban greening benefits for health and well-being across the life-course. In Coles, R. W., Costa, S., Watson, S. (eds). *Pathways to Well-being in Design Examples from the Arts, Humanities and the Built Environment*, Chapter 3. Abingdon: Routledge, pp. 54–75.

English Nature. (2009). *Green Infrastructure Guidance*, Natural England Publications, NE 176. London: Natural England. http://publications.naturalengland.org.uk/publication/35033

Gray, J. (2003). Open spaces and dwelling places: Being at home on hill farms in the Scottish Borders. In Low, S., Lawrence-Zuniga, D. (eds). *The Anthropology of Space and Place: Locating Culture*. Oxford: Blackwell Publishing.

Hassenzahl, M. (2004). The interplay of beauty, goodness, and usability in interactive products, *Human-computer Interaction* 19(4), 319–349. https://doi.org/10.1207/S15327051HCI1904_2

Hassink, J., Van Dijk, M. (2006). *Farming for Health: Green-Care Farming Across Europe and the United States of America* (Vol. 13). Berlin: Springer Science & Business Media.

Hawkins, J. L., Thirlaway, K. J., Backx, K., Clayton, D. A. (2011). Allotment gardening and other leisure activities for stress reduction and healthy aging, *HortTechnology* 21(5), 577–585. https://doi.org/10.21273/HORTTECH.21.5.577

Hewitt, P., Watts, C., Hussey, J., Power, K., Williams, T. (2013). Does a structured gardening programme improve well-being in young-onset dementia? A preliminary study,

British Journal of Occupational Therapy 76(8), 355–361. https://doi.org/10.4276/030802213X13757040168270

Ingold, T. (2000). *The Perception of the Environment: Essays on Livelihood, Dwelling and Skill.* Abingdon: Routledge.

Ingold, T. (2014). A phenomenology with the natural world, *Environmental and Architectural Phenomenology* 25, 22.

Jahncke, H., Hygge, S., Halin, N., Green, A. M., Dimberg, K. (2011). Open-plan office noise: Cognitive performance and restoration, *Journal of Environmental Psychology* 31(4), 373–382. https://doi.org/10.1016/J.JENVP.2011.07.002

Johnson, E. (2018). Alone together-documentary film-making and stories of well-being in outdoor spaces. In Coles, R. W., Costa, S., Watson, S. (eds). *Pathways to Well-being in Design Examples from the Arts, Humanities and the Built Environment.* Abingdon: Routledge, pp. 203–222.

Kaplan, S. (2001). Meditation, restoration, and the management of mental fatigue, *Environment and Behavior* 33(4), 480–506. https://doi.org/10.1177/00139160121973106

Lee, A. C. K., Maheswaran, R. (2011). The health benefits of urban green spaces: A review of the evidence, *Journal of Public Health (Oxford, England)* 33(2), 212–222. https://doi.org/10.1093/PUBMED/FDQ068

Lennon, M., Scott, M. (2014). Delivering ecosystems services via spatial planning? Reviewing the possibilities and implications of a green infrastructure approach, *Town Planning Review* 85(5), 563–588. https://doi.org/10.3828/TPR.2014.35

Massey, D. (1995). The conceptualization of place. In Massey, D., Jess, P. (eds). *A Place in the World?* Oxford: Oxford University Press.

Millman, Z. K. (2013). Photographic postcards as research tools: The postcards from the cut study, *Graduate Journal of Social Science* 54–75.

Morris, G. P., Beck, S. A., Hanlon, P., Robertson, R. (2006). Getting strategic about the environment and health, *Public Health* 120(10), 889–903. https://doi.org/10.1016/J.PUHE.2006.05.022

Natural England. (2010). *Nearby Nature – Accessible Natural Greenspace Guidance, NE 265.* www.ukmaburbanforum.co.uk/docunents/other/nature_nearby.pdf

Pickard, B. R., Daniel, J., Mehaffey, M., Jackson, L. E., Neale, A. (2015). EnviroAtlas: A new geospatial tool to foster ecosystem services science and resource management, *Ecosystem Services* 14, 45–55. https://doi.org/10.1016/J.ECOSER.2015.04.005

Pink, S., Hubbard, P., O'Neill, M., Radley, A. (2010). Walking across disciplines: From ethnography to arts practice, *Visual Studies* 25(1), 1–7. https://doi.org/10.1080/14725861003606670

Rayner, G., Lang, T. (2012). *Ecological Public Health: Reshaping the Conditions for Good Health.* Abingdon: Routledge.

Rogerson, M., Barton, J. Bragg, R., Pretty, J. (2017). *The Health and Wellbeing Impacts of volunteering with the Wildlife Trusts.* University of Essex. www.wildlifetrusts.org/sites/default/files/2018-05/r3_the_health_and_wellbeing_impacts_of_volunteering_with_the_wildlife_trusts_-_university_of_essex_report_3_0.pdf.

Sawang, S., Guaralda, M. (2018). Well-being restoration in the workspace. In Coles, R. W., Costa, S., Watson, S. (eds). *Pathways to Well-being in Design Examples from the Arts, Humanities and the Built Environment.* Abingdon: Routledge, pp. 171–185.

Schmutz, S. U., Lennartsson, M., Williams, S., Devereaux, M., Davies, G. (2014). *The Benefits of Gardening and Food Growing for Health and Wellbeing Health Growing Food Growing for Health and Wellbeing.* London: Garden Organic & Sustain. www.gardenorganic.org.uk/sites/www.gardenorganic.org.uk/files/GrowingHealth_BenefitsReport_0.pdf

Shortt, N. K., Rind, E., Pearce, J., Mitchell, R. (2014). Integrating environmental justice and socioecological models of health to understand population-level physical activity, *Environment and Planning A* 46(6), 1479–1495. https://doi.org/10.1068/A46113

Sitzer, V. A. (2013). *The Effect of a Rejuvenation Room on Reducing Subjective Work-place Stress Reported by Clinical Staff.* San Diego: Sharp Memorial Hospital.

UK Green Building Council. (2015). *Demystifying Green Infrastructure.* www.ukgbc.org/ukgbc-work/demystifying-green-infrastructure/

van den Berg, A. E., van Winsum-Westra, M., de Vries, S., van Dillen, S. M. (2010). Allotment gardening and health: A comparative survey among allotment gardeners and their neighbors without an allotment, *Environmental Health: A Global Access Science Source* 9(1), 1–12. https://doi.org/10.1186/1476-069X-9-74/FIGURES/1

Viljoen, A., Bohn, K. (2009). Continuous productive urban landscape (CPUL): Essential infrastructure and edible ornament, *Open House International* 34(2), 50–60. https://doi.org/10.1108/OHI-02-2009-B0006/FULL/XML

World Health Organisation (2016). *Urban Green Spaces and Health.* Copenhagen: WHO Regional Office for Europe. www.euro.who.int/en/health-topics/environment-and-health.

Zick, C. D., Smith, K. R., Kowaleski-Jones, L., Uno, C., Merrill, B. J. (2013). Harvesting more than vegetables: The potential weight control benefits of community gardening, *American Journal of Public Health* 103(6), 1110. https://doi.org/10.2105/AJPH.2012.301009

4

CHOREOGRAPHY OF THE LANDSCAPE EXPERIENCE

Introduction

> I love this place . . . There is a lot throughout the year and at the moment there are leaves lying on the top of the water, but at other times of the year you get sometimes green coming on it (. . .), is quite varied. Very lush . . . and I also like some of the other plants. There is a particular plant, called a sensitive plant that I always love touching on the knuckle and watching the leaves react or shrivel up. I'm doing that now and it's really wonderful to watch how rapidly it reacts, it's really wonderful. I love the colours of the leaves, the varied colours in the veins and the pinks and reds and greens and shades of green. And the fronds of the leaves, they've just been watered which is rather lovely. So there's drops of water which is super as the sun glimpses through.
>
> — John's journey through the landscape (Costa & Coles, 2018)

This chapter discusses the detail of how people experience landscape places and decipher their complexity and dynamics. In particular, it examines how individuals interact with different landscapes, how interaction occurs and how this influences the state of mind, including well-being. This process is described by us as 'the choreography of landscape experience', a term which we feel explains how the body moves and aligns itself with the environment, the sequencing and speed of movement, actions, flows, and the sensory stimulation involved.

The choreography of the landscape experience thus considers the process of physical and mental 'immersive interaction' associated with an environmental encounter, the connections, flows and mechanisms underpinning such encounters and consequent derived benefits. This includes how these processes are prompted, the spatial and personal constructs concerned, how these are 'of the individual', i.e. relate to the person and their reaction, but also 'of the place', i.e. relate to the

DOI: 10.4324/9781003099758-5

landscape being experienced and its qualities. These concern how the user navigates and interacts spontaneously and organically, and how it is this 'self-choreography' through which actions and steps are composed, arranged, manipulated and negotiated in relation to personal experiences and yearnings, assisted by the intrinsic dynamics of place, its attributes and the context of the experience.

Discussion is underpinned by comprehensive work exploring people's encounters with urban green environments involving planned self-narrated walks with users, focusing on locations in the UK and Portugal. Self-narrated walking prompts individuals to verbalise their walking experience while in the landscape, and essentially involves the action of talking to oneself, where actions, emotions and reactions are brought to the fore. This has proven to be an effective means of understanding personal landscape experiences (see Costa & Coles, 2018; Millman, 2013). In addition, photographs and GPS technology allow the monitoring of movement patterns to relate these narratives to the qualities and nature of the places experienced. Through analysis, it is then possible to identify the choreographies involved, the factors that underpinned their development and how personal strategies showed a sophisticated and personal association with the landscape.

Investigation is thus set in the wider context of more fully understanding the power of landscape systems to engage the individual in ways that are positive and, accordingly, relates to the wider paradigm concerning well-being and environment, but it extends beyond the focus on natural systems and can be considered in terms of the broader functioning of urban green infrastructure.

Choreographies of Place and Nature

> Landscape experience is underpinned by a constant dialogue between person and landscape, in which rhythms of place are superimposed and interweave with the multiplicity of rhythms and actions of the person. Together they form the choreography of landscape experience.

Landscapes are complex environments, which evolve and undergo morphological changes exhibiting visual, spatial and ecological transformation (Olin, 2011). Changes in seasons, the weather and finer detail, such as the reflection of light on water, shadows, textures, colours, breezes and clouds, are aspects which hold people's attention. As such, our environments contain movements and rhythms that form the 'choreography of the place'. These can be intrinsic and perceived in terms of their features, composition, dynamics and fluidity, as well as temporal aspects (for example, the new becoming old, sounds coming and going). Even those aspects that are commonly understood as static, such as pavements or stones, can rapidly change from dry to wet, from cold to hot.

People observe natural processes and landscape changes (Figure 4.1); they appreciate the variety of rhythmical events, they respond emotionally and physically to the aesthetic qualities and attributes and it is the diversity of these processes that has the

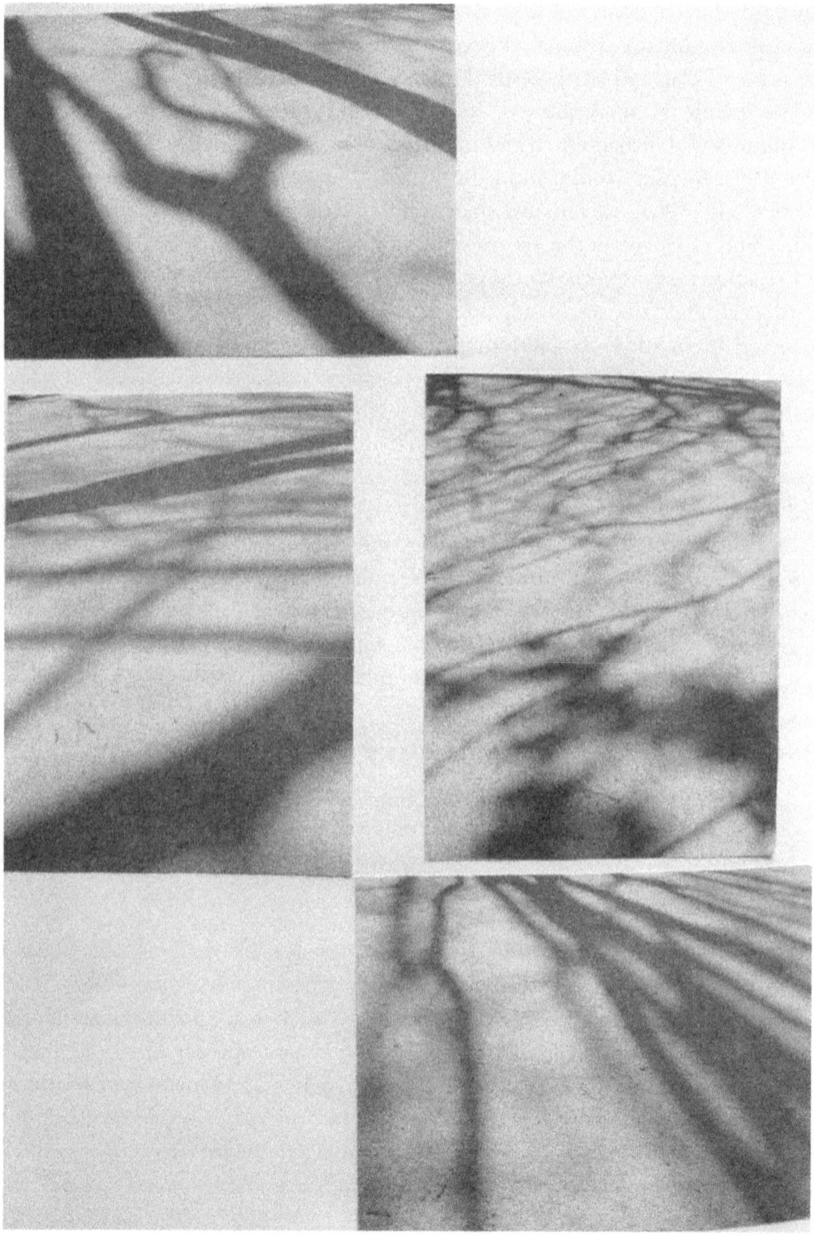

FIGURE 4.1 Capturing the fine detail of a landscape encounter; shadows of tree branches, which make patterns on the ground and recorded in a diary.

FIGURE 4.2 Interaction with art installations and the landscape.

ability to evoke positive states (for example, see Schroeder, 2002). They are drawn by the diversity in vegetation and associated features, the sculptures and installations, (Figure 4.2) water, benches, the topography, the sky, how views are framed and others revealed, but reactions are also established by their wider context – animals grazing, crops growing, the presence of people and their activities, walking, sitting out or children rolling down the lawn. Different sensory cues are experienced, the scent of the roses, the sounds of the city, cars and buses, the sounds of nature, the sounds of water, of a fountain, the 'birds twitting', close or far off, textures of the leaves and the 'crunch' of the gravel. The presence of these changes triggers powerful associations, which extend beyond the moment of the encounter; they make reference to the past and inform new situations in the sense that the place is experienced by a compilation of temporal layers and encounters. It is previous experiences at different times and locations which form a precedent for new interpretations in new circumstances, how these provide a template for assessing the qualities of a new encounter and thus influence the meaning of the experience where people's lives are 'tangled up with the temporalities of constantly unfolding landscapes' (Lund & Benediktsson, 2010, p. 6).

Berleant (2004), for example, stresses the importance of physical presence in the environment and how new perspectives are revealed, how the experience becomes different, changeable and contextual, being dependent on those factors, connections and associations, that combine to create a unique personal, social, cultural and historical situation (Berleant, 1992, 2010).

Negotiating the Experience and Movement to Positivity

Environments which are perceived as satisfying provide a high degree of immersion via sophisticated sensory engagement and in meeting expectations. Essentially, some aspects of the environment are perceived as satisfying and provide a draw, while others are less so. Therefore, the ways that people experience landscape involves negotiation. The process of negotiation is an individual one, although it may be affected by the presence of others. It is led by personal constructs, and is contextual regarding the circumstances of the encounter (see Box 4.1).

BOX 4.1 NEGOTIATION PROCESS

- A negotiation is the process in which an individual tries to achieve an advantageous and satisfying outcome for themselves given the available conditions and possible lines of action and reaction. For example, some environments are perceived as secure and safe, and some are not.
- The process typically involves positive and negative interactions, emotions and reactions, which are associated with self and relate to the specifics of the environment.
- The aim of the negotiation is to find situations which are satisfying and positive, reacting to them and taking necessary action.
- In negotiating the experience, a satisfying experience can either be found in the place where further interaction occurs, by adjusting position or alternatively by moving away to a different location.
- Negotiating the experience allows individuals to move away from negative aspects towards those which they perceive as positive.
- Individuals may not be consciously aware of the factors affecting the process but are able to identify them when prompted.

In exploring the environment, negotiation embraces different actions and reactions which arise from sensory dialogue with the environment as the individual assesses its qualities and relevance to themselves (i.e. assessing its affordances). Broadly, negotiation can be explored within two scenarios, 'compromise' or 'confrontation'.

Compromise – This is the result of a negotiation that balances people's personal constructs regarding what the place being experienced has to offer. It involves assessing both positive and negative aspects but where the final outcome is biased towards the positive, and as such, allows for the person to stay and further explore the qualities of place as a positive place/experience.

Confrontation – Conversely, the result of the negotiation in a particular place may be a confrontation between the place itself and the person; negative factors overwhelm and positive ones recede or fail to materialise. The outcome is that the experience is incompatible with expectations, and the reaction is towards leaving the place and in the search for positivity in another.

To explain the process of negotiation and demonstrating compromise and confrontation, two examples are given. The first case, *compromise*, is supported by the experience of John at the Alpine Garden at the Birmingham Botanical Gardens. The second, *confrontation*, is demonstrated through the experience of Zelen in the same place. The indicator of levels of positivity featured in the case studies are defined taking into consideration the context, the language used to describe experiences and the interaction itself.

John: Case Study 1 – Compromise

When John steps into the Alpine Garden, he proceeds to construct a positive connection. In the beginning, it is immediately acknowledged that his response to the place is positive; words such as 'nice' and 'interesting ideas' demonstrate an indicative level of positivity (modest/low positive). Whilst describing the aspects that are of interest to him in the garden, he also perceives and becomes aware of the negative and intrusive sounds from the road, which momentarily change his perception to a more negative one. However, something in the environment prompts a negotiation between moving away from the intrusive noises and remaining in the place: a personal response to the catmint blue flowers. This moment triggers further involvement and exploration of the interaction, as John decides to sit down, enjoy the colours, take pictures and think about his son.

John's timeline (Figure 4.3) demonstrates that, despite some negative inputs from the environment, the set of positive characteristics arising from his personal realm, i.e. the memory of John's son involved in an episode with the catmint flower, cancelled out the negative ones. In negotiating a better moment for himself, John found a compromise that combined positive aspects self-encapsulated in a memory, allowing negative aspects to retreat into the

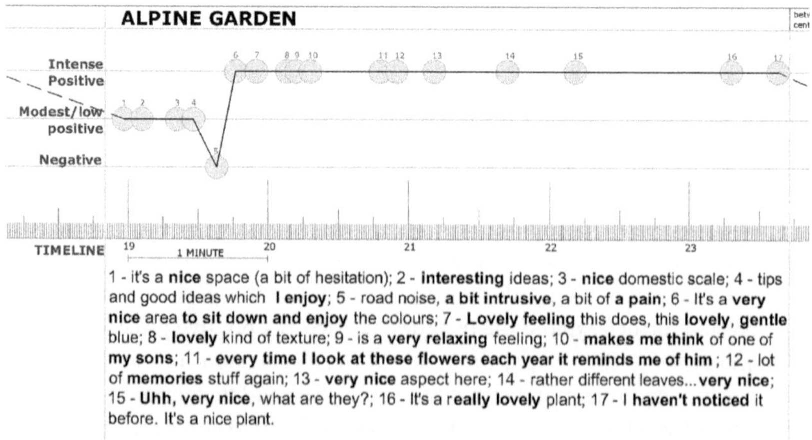

FIGURE 4.3 John's timeline, constructed by using phrases extracted from his narrative which summarise his reactions and emotional state.

background, prompting positive exploration. The experience at this place was a very satisfying and long-lasting emotional moment – a 'very relaxing feeling' and 'each year it reminds me of him', which demonstrates movement to and permanence in a state of positivity.

Zelen: Case Study 2 – Confrontation

Similarly, the timeline for Zelen shows her sequence of events, again at the Alpine Garden (Figure 4.4). Her immediate response was that the alpine features reminded her of mountaineering; she is a keen mountaineer, so in the first instance, there is a positive emotional response and reaction to the place (possibly prompted by the name of the place itself). However, her memory of good moments being in the mountains started to blend with a negative memory of her last mountaineering experience, where she felt emotionally insecure and scared. From this moment onward, a downward spiral of negative emotions and reactions take place. As she walks through, she becomes aware of the physical characteristics of the space, which, despite being agreeable, are disappointing to the extent they are not in any way similar to what she normally finds in the mountains – 'but I just don't get why [this is] the Alpine Garden, because this place doesn't remind me [of] what I find in mountains'. Similarly to John, she notices the traffic from the road nearby and becomes aware of her non-identification with the place, which was the final stimulus for her to leave. In addition to the few negative inputs of the environment (the traffic noise from the road), and the negative memory associations (the memory of a bad experience in the mountains), the place did not respond to expectations regarding alpine vegetation. Little time was spent in the place and disappointment is registered.

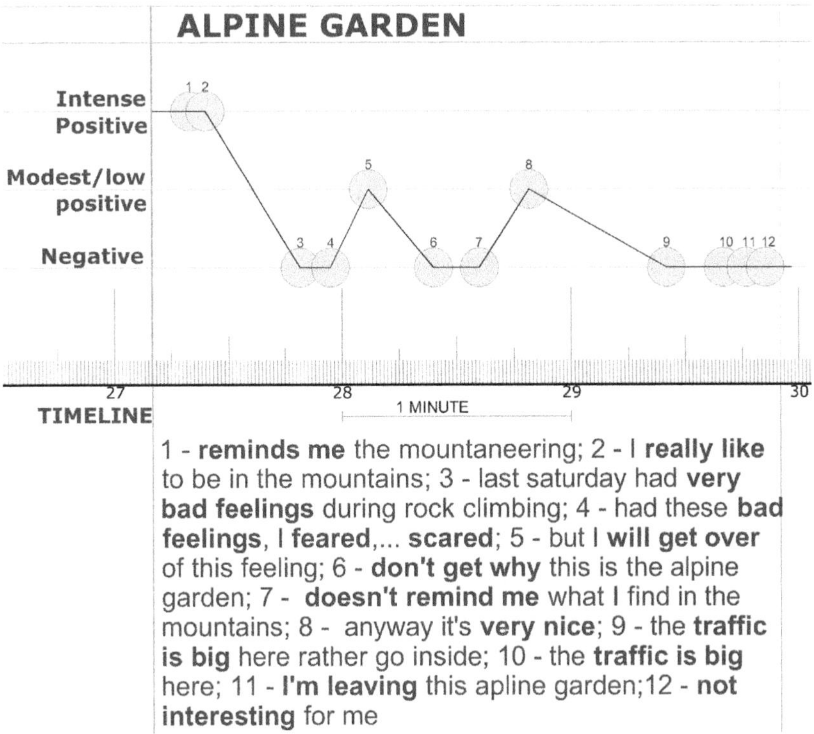

FIGURE 4.4 Zelen's timeline, constructed by using phrases extracted from her narrative which summarise her reactions and emotional state.

What is revealed is a positive, but unstable, moment (when she first approaches the Alpine Garden) that deteriorates and to the extent of disappointment. Zelen is then motivated to move away towards a more positive situation – 'so I rather go inside the garden and be in a more silent space'. This reluctance to remain in the Alpine Garden demonstrates that Zelen did not dwell on the negative and negotiated the experience by moving away and exploring a different location.

Negative Intrusions

As demonstrated by John and Zelen, negative intrusions affect interaction and influence negotiation. They may arise from sources outside or inside the location, or be related to personal negative past events.

- *External* intrusions are originated from outside the place, such as bad weather conditions or intrusive views. For example, some external intrusions raise awareness of the background noises of the city which are not expected in a garden or a park, generating an unwanted connection with the outside world, and demonstrating that the experience is incongruous with or does not meet expectations (traffic and other city noises are a common and ubiquitous negative intrusion).

- *Inherent* intrusions have their source within the environment itself, such as crowded and noisy places, poor or lack of maintenance, absence of certain features, the lack of light and other people's attitudes and behaviours. It also includes situations where place characteristics became incongruous with memories of previous visits, a view that is no longer there, or places that are unexpectedly crowded and noisy and no longer satisfy solitude and reflection.
- *Person-based* intrusions are specific to the individual and are personal, based on negative past events and related concerns, including fear of attack or fear of getting lost.

Positive Emotions

Operating in an opposite way to negative intrusions, positive emotions emerge from personal constructs and from goals, motivations, expectations and attachments that people hold about places (Ward Thompson et al., 2004; Milligan & Bingley, 2007; Aspinall, 2010; Scott & Canter, 1997; Tuan, 1980). Emotions and reactions and intrusions operate at different scales associated with the quality of the experience:

On the lowest spectrum, there are a group of emotions and interactions that, even though they are positive and important to the overall experience, they do not engender high opportunities for deep engagement. Negative intrusions stop the experience progressing to a more positive association.

In this spectrum, we can depict a place that has some positive qualities and elements that are appreciated, but that negative intrusions (associated with either the place or person-based) are too dominant, resulting in an overall sense that the full conditions for a wholly satisfying experience are not demonstrated. This means:

- Place and person are not fully synchronized; there is some degree of discontent.
- People do not linger and move through the space fairly quickly.
- The language used to describe the place is weak.
- The properties of the place are not fully explored.
- The place encounter does not fully embrace expectations; the level of congruence is weak.

On the higher spectrum of positivity, we have indicators of individuals' diversity of feelings and positive situations that combine the positive personal constructions and the positive features present in the environment. Any negative intrusions are trivial and hardly noticed. Place and person are in harmony, and there is congruence between past experience and expectations. The result of these transactions leads to:

- Slowing down and lingering – adopting stationary positions (e.g. sitting and standing still)
- Full exploration of the place, involving greater interaction with the landscape associated with immersive and introspective actions.
- A high level of emotional and physical engagement, including the recollection of happy memories and events.
- Silent moments, moments of reverie.
- Relaxation and restoring mental fatigue.
- A high level of close interaction and purposeful engagement with the landscape (including, touching, smelling and listening), being more aware, more mindful and open to its qualities.
- Associations and positive comparisons with other favourite special places, but also appreciation for rare and unexpected events.
- Moments associated with aspects of fascination, pleasure and a sense of wonder.

The narratives in our research show that an experience is highly positive when a great number of intense positive expressions prevail, i.e. when the result of the negotiation is highly favourable to the person, and as a result there is permanence in the place and reluctance in leaving it. Notwithstanding this, it is believed that interactions and emotions at the lowest spectrum of interaction are important in the construction of more intense positive ones. That is, by contrast and comparison, they tend to give greater strength. Part of the choreography is experiencing both negative and positive, weighing up impacts, assessing these in the light of personal experience, relevance and taking action, either to linger and enjoy the moment or to move on.

Negotiation appears to be part of the learning process regarding the landscape, involving exploring and understanding the environment from the perspective of self. This is a process through which the individual is able to secure a positive experience, if necessary to adjust their perspective by moving to a more preferable location. Through negotiation, people align themselves, their memories and expectations to personal advantage, realising and enacting their personal choreographies.

The ability or freedom to construct and live positive emotions when experiencing the landscape would seem critical to environmental performance. People negotiate the landscape and achieve moments of deep immersion. These involve positive emotional states such as joy, peace, satisfaction, pleasure, inspiration, curiosity or love (Seligman, 2012), and the level of positivity that individuals seem to attain when they are in a high state of joy and full involvement with the environment is congruent with a 'flow experience' as described by Csikszentmihalyi et al. (2005). Other studies have found that pleasure, engagement and meaning correlate with well-being and underpin happiness (Schueller & Seligman, 2010).

In this context, in terms of the reality of the urban landscape, it appears that much of the general experience is biased towards the lower end of the spectrum (there are many negative intrusions in the wider urban scene), and the environment does not have the capacity to fully immerse the user. While negotiation might be expected and thus a move to situations that do provide appropriate immersion, it is questionable if there are sufficient opportunities to do so: they may be too remote (too far away) to offer the immediacy required, or not sufficiently robust to maintain it. Even in places where the conditions are 'right', for examples parks, negotiation takes place in the choice of a favourite location or route.

Choreographies and the Walking Speed

Walking is a very comprehensive experience that includes different paces, stopping and sitting, as well other associated movement patterns. The speed of walking affects the experience, moderates interaction and thus perception; conversely, landscape quality affects the pace of movement. Speed and pace produce dynamic ways of knowing and feeling spaces and places, and this shapes an experience that is distinctive to an individual (Edensor, 2010a). In situations where people have freedom and option to control their own actions and time, the experience can be understood by observing the pace of movement in relation to the landscape dynamics involved; variation in pace from hurried to stationary is an integral part of choreographies.

In our study, individuals showed a pace lower than the average of speed of a normal walk (Knoblauch et al., 1996). When we combined GPS information of people's walks with their verbal narratives (Costa & Coles, 2018), many of the highly positive interactions occurred precisely when individuals slowed down or stopped and the places where they stopped and spent more time were highly related to their favourite places, special places or with another sort of attachment to place.

Gentle movements, slower encounters and pauses allow people to take in the surroundings and identify further opportunities for stopping and sitting down. They empower individuals to look closer into the landscape, to interact and to notice its complexity. These 'gentle' encounters provide moments of spatial awareness and intimacy with the environment, and can be linked with flow experiences of immersion and reflection (Nakamura & Csikszentmihalyi, 2009; Csikszentmihalyi et al., 2005).

Studies have demonstrated that the pace can vary depending on the characteristics of the place and on personal factors, including terrain and health (Knoblauch et al., 1996), but having things to see and interact with affects pace. Slowing down gives people opportunities to interact, e.g. stopping temporarily to eat, to draw, to take pictures, to talk to someone, to contemplate, observe the landscape or to engage in haptic and other bodily gestures. Conversely, we were able to observe that individuals' accelerations during the self-narrated walks tended to produce vaguer accounts with the regards to their perceptions and the details of the landscape (see Box 4.2).

BOX 4.2 HOW THE WALKING SPEED AFFECTS INTERACTION

- Spatial perception decreases with the increase of speed, or, in other words, as speed increases, space perception declines, as well as the understanding of place.
- Greater interaction with the landscape occurs when individuals slow down or stop.
- However, accelerations, i.e. moving through the landscape quickly, play a significant part in the choreography by contributing to the unfolding of the landscape and sequencing of places, developing expectations and curiosity to explore something that lies ahead or around the corner.
- The purpose of the walk and self-identification with the place are factors that influence a slower pace.
- Overall, the pace of movement reflects patterns of engagement, and its variation is part of the dynamics of landscape choreography.
- Pace, movement and landscape are interrelated and concern different modes of engagement or disengagement.

Research seems to suggest that the tendency to slow down and to stop in particular places and to speed up in others are affected by a number of factors. Some places, through their intrinsic characteristics, will have a tendency to convey repose and quietness associated with slowing down and stopping, whereas others will evoke movement. Places have their own 'atmosphere' which influence people's pace (Löw, 2008). There is also evidence that pace and speed are dependent on the intention of the walk. Different walking patterns are associated with different tasks, i.e. different intentions or modes of engagement, and they produce distinct rhythms of walking (Lefebvre, 2004; Edensor, 2010). The strolling tourists in the city and a meandering stroll in the park show a leisurely kind of intention, associated with a slower pace; other situations may not, or may send mixed messages.

The Multiplicity of Interactions

People explore the environment through self-identity and through the ways that the body aligns itself with the environment seeking to maximise the experience and fulfil its potential. The environment is thus related to the representation and exploration of self, reflected through the ways that the body is immersed in it and the ways in which the environment accommodates immersion. It is this multiplicity of interactions that is contained within the choreographies of landscape, place and person, and which allow for a positive experience underpinned through the characteristics, qualities and dynamics of the environment, movement and negotiation. The broader processes and practices involved are summarised below:

- Movement and negotiation are underpinned by sensory engagement; negotiation involves movement to optimise sensory engagement in the context of self (Wunderlich, 2008; Ingold, 2004).
- The perception of qualities of place involve two particular kinds of interaction: participative, being affected by the qualities of space, and introspective, concerning the individual.
- There are synergies between the individual and the environment; places are interactive and dynamic, and in each encounter, people explore the duality of this relationship.
- This multiplicity of interactions are exhibited and rehearsed in individual choreographies, and can be analysed further in relation to environmental quality.
- The physical environment will determine and affect key activities and affords opportunities for interaction.
- The potential of environments is assessed in terms of their affordances, i.e. assessed regarding their ability to support actions, including realising positive emotional states.
- Environmental features underpin reactions and actions in realising affordances and have an effect on behaviour.
- Interactions are affected by the personal intent behind them.

The multiplicity of interactions in which people pursue personal enjoyment, amusement, relaxation and introspection are multifaceted, but many are nature based (see Chapter 2). Within this multiplicity, there is a range of detailed processes occurring, including how the different senses are stimulated and how they reinforce the experience (how the experience is lived).

Haptic Connections

Environmental engagement is evoked through the sense of touch, e.g. touching plants and feeling textures (Figure 4.5). As such, touch is part of the process involved in the deep exploration of the environment, and a desire to touch is seen in situations where deeper immersion is occurring. Haptic experiences represent purposeful interactions that are characterized by exploration and responsive contact with the environment involving environment-based stimulation. The depth of the interaction will depend on how the senses are engaging and to what level they are being heightened by 'haptic sensibilities' (e.g. touching and being touched), and serve to realise the affordances, to deepen and underpin the experience. Many authors have made mention of touch and haptic sensations as ways for perceiving the landscape and moderating the interface between self and the landscape (Ingold, 2004; Lund, 2005; Wylie, 2005, 2009; Pallasmaa, 2005). In our study, haptic interactions were associated

FIGURE 4.5 Touching is associated with deeper environmental engagement. These images demonstrate the choreography of the experience in nature places including haptics. Touching leaves; interacting with water (throwing a stick); drawing closer and touching tree bark; the texture under foot; kicking and walking on autumn leaves; the distinctive smell of fungi and autumn.

with highly positive narratives, and equate with a heightened sensory experience which contributes to generating proximity between person and place or person and nature, creating affinity and intimacy, a higher level of engagement and is associated with the movement to a more positive state and therapeutic feelings.

Silences

Silences carry meaningful and purposeful interaction and are integral to the choreography of the experience. Silent moments can be difficult to interpret, and in qualitative analysis they are often understood as the lack of communication. However, silence is a purposeful reaction to a situation and a reaction inherent in choreographies where words fail to express feelings and where the qualities of the surroundings enforce silences. Silence is required to explore, activate or to awake the senses, and silence (or quietude) is seen as an important quality in the appreciation of nature. Silence seems to trigger sensory experiences by encouraging the senses to open up to environmental stimuli. Similarly, silence is linked to self-introspection, reflection and self-reorganisation – aspects considered therapeutic, being associated with feelings of relaxation and with transcendent experiences of tranquillity, peace and meditation (Tuan, 1977; Wilson, 1984; Ehrenhaus, 1988; Kaplan & Kaplan, 1989). For example:

> It feels good to stay like this in silence, silence is good. Enjoy the break between our varied thoughts . . . thinking of nothing is also good. We can contemplate, feel the breeze, feel the nature living and photograph this moment.
>
> *[Lara, self-narrated walk]*

> I sit in the sun, I close my eyes . . . I seek the warmth. I hear a woodpecker . . . I feel good. I wish to be and to stay.
>
> *[Lara, diary]*

> This is a perfect space. I think that even birds can be heard . . . It is a space where you can hear very well . . . the fact of being with someone else, even if in silence, is comforting.
>
> *[Lara, self-narrated walk]*

Memories

Memories of other landscapes operate as a 'placing mechanism' in which values of other landscapes are assigned to those actually being experienced (Oliveira, 2011; Tolia-Kelly, 2004). One of the most powerful themes arising from researching people walking and experiencing the environment is related to memory associations and the ability the environment has to evoke memories. Sights, sounds, smells, people, conversations and events of childhood and adulthood are remembered through association with places that are known and familiar (Figure 4.7). These place-images are

FIGURE 4.6 Strategically located benches overlooking the lake at Serralves Park, Portugal, one of its users' favourite places.

memory triggers that inform perceptions and meanings of landscapes and become mediators through which present experiences are constructed. These sensory recollections 'contextualise our body experiences within past experiences of dwelling, environment and places of being'. They are a powerful means to construct a sense of self (Tolia-Kelly, 2004, p. 287), and contribute to people making sense of the experience (Millman, 2013; Degen & Rose, 2012; Jones, 2010). When people recall places and experiences, it means that they are not only engaging with the landscape as it exists, but also with how it was (Degen & Rose, 2012).

Memories have different associations or contexts:

Memories associated with time spent in the environments being experienced. Often associated with particular memories of individuals themselves and/or of being with other people of their circle, their family and their children, playing and enjoying the place. This may include reliving previous encounters in the same or similar environments, or reinforcing these by recalling past situations. Recent and/or old events which were retained in their minds and their hearts are recounted and reinforced with great detail, enthusiasm, emotion and nostalgia. Childhood memories can be especially powerful.

Memories of other environments. Triggered by the features of the environment, some places evoke memories and transport us to elsewhere, other places, other landscapes. They evoke memories from holidays or places we have been, countries where we lived before or places where we were brought up. They are activated while moving through the environment and triggered by different cues, (Tolia-Kelly, 2004).

Searching for pertinent memories and the pursuit of movements to other places and times. Memories of past places are used to interpret new ones and new encounters may be rationalised (or decoded) by referring to familiar situations even though they are far removed from the reality of the current encounter. It occurs when individuals search inside and around themselves and the environment for the memories, for example, Amira's narrative: 'this part reminds me Bangladesh, it smells like Bangladesh, it feels like Bangladesh . . . even the fish . . . and the banana trees . . . the sounds. Every time I come here, I feel as I'm in Bangladesh actually.' (Amira, self-narrated walk).

The physicality of the memory. Remembering encourages physical interaction with places (Figure 4.7). It expresses things that a person would have done in the past. Remembering goes beyond the mental representation; it is transferred to the body into a muscular movement and bodily actions associated, and concerns the physical reconstruction and reinterpretation of past experiences; the body remembers through haptic connections (Pallasmaa, 2005). Indeed, authors in the field of movement therapy who explore the concept of 'body-mind' and 'embodiment' suggest that the body accumulates emotional history in the muscles and other physiological systems (Chaiklin & Wengrower, 2010) and thus that memories are stimulated through certain ritual physical actions.

Favourite Places, Place Attachment and Affinity

In terms of experiencing the landscape, favourite and special places are those that have the power to evoke strong interactions between self and environment and demonstrate the multiplicity of interactions discussed. Favourite places demonstrate

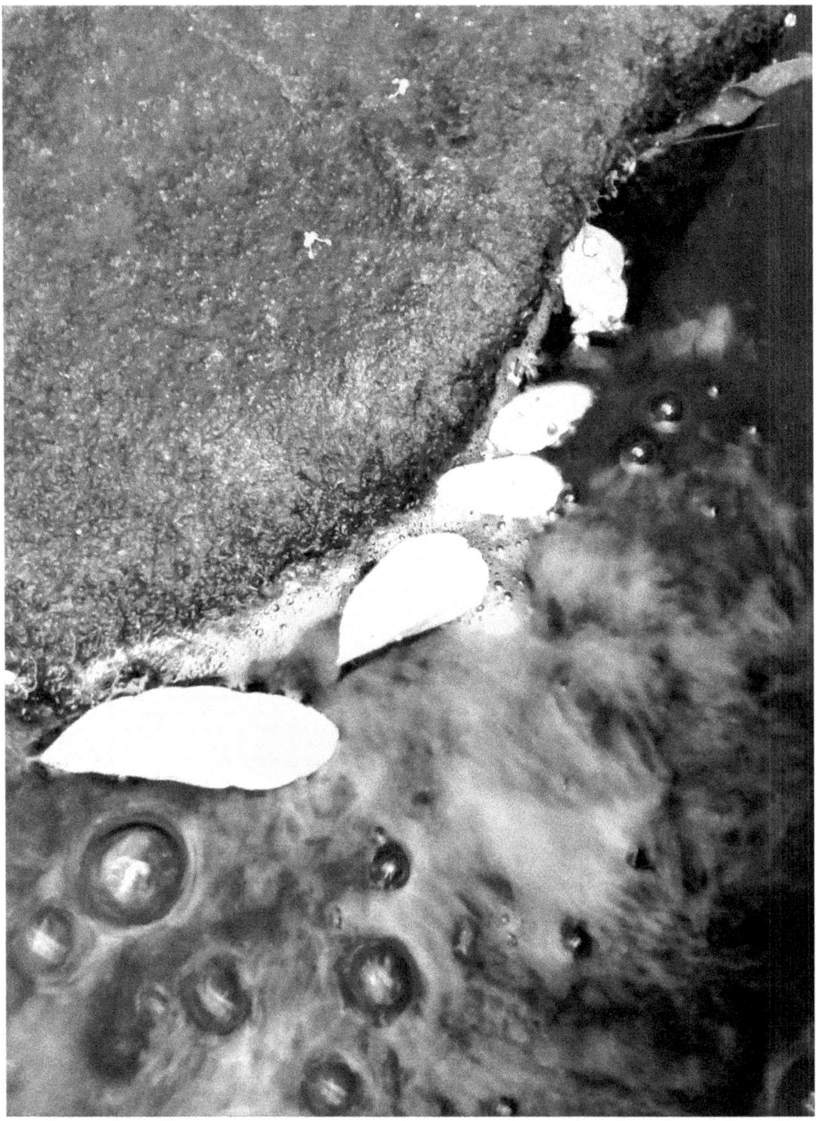

FIGURE 4.7 The physicality of the memory. Reconstruction of a past experience – running leaves in the water whilst recalling memories of a childhood landscape (Fontainebleau) where the participant used to have a similar experience.

high degrees of positive interaction, often expressed in strong words and actions, but underpinned by reference to past events, to memories and to features, all combined to describe its attributes (Box 4.3). They demonstrate lived experiences which address the present, the past, values and encourage related actions, deep immersion and typically a sense of reverie.

- Favourite places relate to places of proximity, familiarity and of affinity, often places associated with childhood.
- Favourite places encompass the duality of place and personal identity evoked by an experiential mechanism that combine the spatial, aesthetic and sensory qualities together with what a person brings to it.
- Favourite places engender intense positive transactions, and there is a positive correspondence between place and person.
- Preference for favourite places is related to personality and the personal realm.
- Favourite places are associated with feelings of peace, relaxation, tranquillity, serenity, quietness, comfort and losing self.
- Favourite places are associated with positive and sensory place qualities, and positive emotional narratives (Box 4.3), where expressions include 'secretive', 'dreamy', 'lovely big plants', 'beautiful colours', 'lovely contrast', 'sound of running water' and 'pleasant summer warmth'.
- Favourite places involve choreographies that are personal constructs where responses are influenced and amplified, i.e. made more profound, by personal values, tendencies and introspection, being reinforced by repeated experiences and congruent memories.

The diversity and aesthetic quality of features and the sensory attributes of place are aspects that contribute to make a place attractive and positive. These places and their attributes prompt moments of joyfulness and happiness, peacefulness and relaxation, as well as feelings of safety, mystery and intimacy. Seasonality

BOX 4.3 TWO EXAMPLE NARRATIVES ASSOCIATED WITH FAVOURITE PLACES

> And part of the reason for it [the cottage garden] being a favourite spot is that it's sort of like a secret garden . . . all these lovely big plants . . . it's cottagey and I can just sit and lose myself in here. Beautiful colours the leaves of the trees . . . There's these beautiful campanulas and foxgloves and I think that combination is beautiful . . . and the shape of the flowers are similar and yet that's a lovely contrast, which I think is great.
>
> [John, self-narrated walk]

> This always has been one of my favourite places, the rock garden. Again I think it reminds me [of] my dad's garden where he had waterfalls and a lovely rock garden . . . and just the sound of that running water is really peaceful.
>
> [Joanne, self-narrated walk]

FIGURE 4.8 Special places. Special memory benches celebrate both the view and the person (Birmingham Botanical Gardens).

and landscape changes bring diversity, novelty and renewal to experience; different sounds, smells and sights all have an important role in enhancing sensory qualities. Favourite and special places are assigned with value and significance because they are constructed and shaped in relation to life experiences and personal expectations. As such, they involve strong emotional attachment, relate to cultural identity, reflective experiences or wider social engagement (Figures 4.8–4.10). They offer opportunities for transcendent and meditative experiences, as well as allowing 'self-regulation' e.g. after a negative event (Coles & Millman, 2013; Korpela et al., 2020, 2008).

Conclusion and the Wider Urban Context

This chapter has presented ideas regarding how people interface with the environment where interaction is referred to as the choreography of the landscape experience, a term which describes the nature and purpose of engagement that expresses a form of a dialogue between people and place, and which is scripted by sensory responses, innate biophilic connections and personal factors. Findings are identified through the use of self-recorded narratives where the participant essentially talks to themselves, and in doing so, is able to verbalise the thought processes and the related environmental cues that underpin the experience. Analysis of these

FIGURE 4.9 Favourite places. Favourite places are constructed according to expectations for use and for drawing personal benefits, resulting from interactions. A lawn will offer opportunities for both social interaction, such as picnicking with the family or listening to a band play on Sundays, and for solitary moments of sitting and contemplating the landscape (Stratford-Upon-Avon).

narratives provides ample evidence of the sophistication and importance of individual choreographies, including the emotional and environmental contexts which are involved.

Findings indicate that people move through the environment continuously examining its qualities, and react to sensory cues (although they may not be overtly aware of the process) by seeking out situations that place them in a favourable emotional state. Thus, the driver is emotional and physical satisfaction, which occurs when there is empathy between the environmental qualities and the person's requirements. The process is bounded by individual responses, which are themselves the product of past experiences and associated memories. What is demonstrated is that, in the right circumstances, there is a resonance between place and person which is reflected in a crescendo of positive emotions underpinned by the excitement of the senses.

Accordingly, interaction is not a neutral process; the environment is interpreted, ratified and consolidated in the context of innate reactions to nature and the natural world and past experience. This is a situation explained as negotiation, where different environmental qualities are considered in terms of negative intrusions

FIGURE 4.10 A favourite place at Serralves Park. The journey down to the lake alongside established planting, cascades and stopping points for sitting and contemplation. Enclosed areas with water are sought due to the atmosphere and qualities which offer opportunities for more introspective and transcendent experiences/interactions and for moments of silence. When expectations are met, places can deliver highly positive experiences and contribute to more positive states of being.

and positive emotions. As a result, negative and positive aspects are automatically assessed. A negotiation is successful when positive factors overcome any negative factors and decipher the qualities of the experience as a positive one that meets need, and then prompt deeper exploration of the situation, both in terms of the environmental qualities and personal connections. It is in these situations that moments of high positivity and meaningful emotional intensity can occur. It is then that a purposeful exploration of the finer qualities of the encounter drives the experience into a deeper form of engagement and where it reaches its peak with positive effects and the initiation of a positive flow experience. In situations where there are too many negative intrusions, then there is no drive to explore it further.

While the process might seem rather complex and procedural, it is believed to occur spontaneously, but is brought to a more conscious level through the act of self-narration, where the self-narration process encourages a more mindful and reflective approach to engagement. This involves a heightened awareness of both self and environment, and is a technique that can be adopted to advantage in other situations as a mechanism for amplifying an awareness of environment. The importance of being able to access landscapes/environments which endorse personal choreographies is emphasised, so that people can negotiate their individual experiences and place themselves in different landscape situations to advantage and as a result, experience the associated restful, spiritual and emotional advantages.

In the places studied, the landscapes were set aside for recreational purposes and scientific study, containing sub-environments which were nature based, including native vegetation systems and representations of nature, plus ornamental gardens. In these conditions, negotiation was a relatively simple task and favourite 'peak performance' locations easily assessed. Similar situations are present throughout the wider urban environment, but it is probable that many general local environments, at least in the ways that they are currently presented or are accessed, do not have the resilience or composition to allow effective negotiation.

In addressing urban landscape futures, it is clear that, as a guiding principle, environments must offer scope to be explored by walking, offer characteristics that promote different pace/movement and be of a quality that allows negotiation to take place. Individuals can then explore the landscape on their own terms and make sense of it by being able to relate their own experience to the various cues and features that it contains. Green infrastructure can then be strengthened as required (Box 4.4). It is clear that landscapes need to be sufficiently flexible and accessible

BOX 4.4 SUMMARY OF THE KEY PARAMETERS INVOLVED IN THE DEVELOPMENT OF EFFECTIVE LANDSCAPE CHOREOGRAPHIES

- In experiencing the landscape, there are positive and negative interactions, associations and emotions cued by different senses and rationalised by memories of other places.
- Attachment to a place is associated with 'seeing' and realising opportunities for personal benefits.
- Individuals negotiate the experiences which best support positivity through an open and dynamic dialogue involving self-identity, environmental context and experience.
- Places prompt the recall of memories of other places and other events; favourite places are associated with personal identity and strong memories linked to places or features in places.
- Effective and favourite places deliver highly positive experiences and contribute to more positive states of being.
- The walking speed in green environments is significantly slower when compared with the speed of a normal walk.
- Slower speeds allow for taking in the surroundings and for closer interaction.

- The presence and diversity of vegetation, through its natural properties, will induce a slower pace promoting deeper engagement.
- Silences are a purposeful part of interaction, and silence and quietness, including the absence of intrusive noise, is a sought-after quality.
- Individual memories, including innate biophilic memory, are integral to the experience and provide the context for interpretation.
- In situations that encourage deeper engagement, interaction involves more intense bodily engagement.
- It is important that the urban environment has the qualities, choice and flexibility to allow people to realise and refine their personal choreographies.

for this to happen. Accordingly, ideas can be considered more generally concerning the lived experiences of urban places, especially regarding the properties of green infrastructure, which contain many features thought capable of supporting personal choreographies, but where the full potential is not actually realised.

Given the importance of memory, specifically space memory, there are questions about the ability of people to more generally fully utilise their local landscapes. Concerns have been raised about the lack of contact with the natural world, i.e. the reduced opportunities to experience nature and thus fewer opportunities for establishing precedent memories of places, or being able to relate the experiences they have to different local situations. Findings suggest that in such a situation, individuals may never develop the required skills of negotiation, refinement of exploration and positive emotional experience that might be available – a situation that needs to be addressed through a range of support mechanisms to ensure that the local landscape delivers to its full potential.

References

Aspinall, P. (2010). On environmental preference: Applying conjoint analysis to visiting parks and buying houses. In Ward Thompson, C., Aspinall, P., Bell, S. (eds). *Innovative Approaches to Researching Landscape and Health: Open Space, People Space 2*. London/New York: Routledge, pp. 179–208.

Berleant, A. (1992). *The Aesthetics of Environment*. Philadelphia: Temple University Press.

Berleant, A. (2004). The aesthetic in place. In Menin, S. (ed). *Constructing Place: Mind and the Matter of Place-Making*. London/New York: Routledge, pp. 41–54.

Berleant, A. (2010). *Sensibility and Sense: The Aesthetic Transformation of the Human World*. Cambridge: Imprint Academic.

Chaiklin, S., Wengrower, H. (2010). *The Art and Science of Dance/Movement Therapy: Life is Dance*. New York: Routledge.

Coles, R., Millman, Z. K. (2013). Landscape, well-being and environment. In Coles, R., Millman, Z. K. (eds). *Landscape, Well-Being and Environment*. Oxon: Routledge, pp. 200–217.

Costa, S., Coles, R. (2018). The self-narrated walk. A user-led method to research people's experiences in urban landscapes, *Landscape Research* 1–15. https://doi.org/10.1080/01426397.2018.1467004

Csikszentmihalyi, M., Abuhamdeh, S., Nakamura, J. (2005). Flow. In Elliot, A. J., Dweck, C. S. (eds). *Handbook of Competence and Motivation*. London: Guilford Publications, pp. 598–608.

Degen, M. M., Rose, G. (2012). The sensory experiencing of urban design: The role of walking and perceptual memory, *Urban Studies* 49(15), 3271–3287. https://doi.org/10.1177/0042098012440463

Edensor, T. (2010). Introduction: Thinking about rhythm and space. In Edensor, T. (ed). *Geographies of Rhythm Nature, Place, Mobilities and Bodies*. New York: Routledge, pp. 1–18.

Edensor, T. (2010a). Walking in rhythms: Place, regulation, style and the flow of experience, *Visual Studies* 25(1), 69–79. https://doi.org/10.1080/14725861003606902

Ehrenhaus, P. (1988). Silence and symbolic expression, *Communication Monographs* 55(1), 41–57. https://doi.org/10.1080/03637758809376157

Ingold, T. (2004). Culture on the ground: The world perceived through the feet, *Journal of Material Culture* 9(3), 315–340. https://doi.org/10.1177/1359183504046896

Jones, O. (2010). Endlessly revisited and forever gone: On memory, reverie and emotional imagination in doing children's geographies. An "Addendum" to "To go back up the side hill": Memories, imaginations and reveries of childhood by Chris Philo, *Children's Geographies* 1(1), 25–36. https://doi.org/10.1080/14733280302185

Kaplan, R., Kaplan, S. (1989). *The Experience of Nature: A Psychological Perspective*. Cambridge: Cambridge University Press.

Knoblauch, R. L., Pietrucha, M. T., Nitzburg, M. (1996). Field studies of pedestrian walking speed and start-up time, *Transportation Research Record* 1538(1), 27–38. https://doi.org/10.1177/0361198196153800104

Korpela, K. M., Korhonen, M., Nummi, T., Martos, T., Sallay, V. (2020). Environmental self-regulation in favourite places of Finnish and Hungarian adults, *Journal of Environmental Psychology* 67, 101384. https://doi.org/10.1016/J.JENVP.2019.101384

Korpela, K. M., Ylén, M., Tyrväinen, L., Silvennoinen, H. (2008). Determinants of restorative experiences in everyday favorite places, *Health & Place* 14(4), 636–652. https://doi.org/10.1016/J.HEALTHPLACE.2007.10.008

Lefebvre, H. (2004). *Rhythmanalysis: Space, Time and Everyday Life*. London/New York: Continuum.

Löw, M. (2008). The constitution of space: The structuration of spaces through the simultaneity of effect and perception, *European Journal of Social Theory* 11(1), 25–49. https://doi.org/10.1177/1368431007085286

Lund, K. (2005). Seeing in motion and the touching eye: Walking over Scotland's Mountains, *Etnofoor* 18(1), 27–42.

Lund, K., Benediktsson, K. (2010). Introduction: Starting a conversation with landscape. In Benediktsson, K., Lund, K. A. (eds). *Conversations with Landscape*. Cornwall: Ashgate.

Milligan, C., Bingley, A. (2007). Restorative places or scary spaces? The impact of woodland on the mental well-being of young adults, *Health & Place* 13(4), 799–811. https://doi.org/10.1016/j.healthplace.2007.01.005

Millman, Z. K. (2013). Photographic postcards as research tools: The 'postcards from the cut' study, *Graduate Journal of Social Science* 10(2), 54–75.

Nakamura, J., Csikszentmihalyi, M. (2009). Flow theory and research. In Snyder, C. R., Lopez, S. J. (eds). *Oxford Handbook of Positive Psychology*. Oxford: Oxford University Press, pp. 195–206.

Olin, L. (2011). Form, meaning, and expression in landscape architecture. In Treib, M. (ed). *Meaning in Landscape Architecture and Gardens*. New York: Routledge, pp. 22–71.

Oliveira, E. S. (2011). *Immigrants and Public Open Spaces: Attitudes, Preferences and Uses*, (PhD Thesis). Edinburgh: The University of Edinburgh.

Pallasmaa, J. (2005). *The Eyes of the Skin. Architecture and the Senses. Architecture*. Cornwall: John Wiley & Sons.

Schroeder, H. W. (2002). Experiencing nature in special places: Surveys in the north-central region, *Journal of Forestry* 100(5), 7.

Schueller, S. M., Seligman, M. E. P. (2010). Pursuit of pleasure, engagement, and meaning: Relationships to subjective and objective measures of well-being. *The Journal of Positive Psychology* 5(4), 253–263. https://doi.org/10.1080/17439761003794130

Scott, M. J., Canter, D. V. (1997). Picture or place? A multiple sorting study of landscape, *Journal of Environmental Psychology* 17(4), 263–281. https://doi.org/10.1006/JEVP.1997.0068

Seligman, M. E. P. (2012). *Flourish: A Visionary New Understanding of Happiness and Well-being*. New York: Simon and Schuster.

Tolia-Kelly, D. P. (2004). Landscape, race and memory: Biographical mapping of the routes of British Asian landscape values, *Landscape Research* 29(3), 277–292.

Tuan, Y. F. (1977). *Space and Place: The Perspective of Experience, Notes*. Minneapolis: University of Minnesota Press.

Tuan, Y. F. (1980). *Landscapes of Fear*. Oxford: Basil Blackwell.

Ward Thompson, C., Aspinall, P., Bell, S., Findlay, C. (2004). *Open Space and Social Inclusion: Local Woodland Use in Central Scotland*. Edinburgh: Forestry Commission.

Wilson, E. O. (1984). *Biophilia*. Cambridge/London: Harvard University Press.

Wunderlich, F. M. (2008). Walking and rhythmicity: Sensing urban space, *Journal of Urban Design* 13(1), 125–139. https://doi.org/10.1080/13574800701803472

Wylie, J. (2005). A single day's walking: Narrating self and landscape on the South West Coast Path, *Transactions of the Institute of British Geographers* 30(2), 234–247. https://doi.org/10.1111/j.1475-5661.2005.00163.x

Wylie, J. (2009). Landscape, absence and the geographies of love, *Transactions of the Institute of British Geographers* 34(3), 275–289. https://doi.org/10.1111/j.1475-5661.2009.00351.x

5

BIOPHILIA AS AN EMBODIED EXPERIENCE

The Emotional Dimension and Agency of the Arts

Introduction

This chapter specifically attempts to widen the understanding of biophilia and associated landscape encounters, and how the experience is an embodied one for the individual as both emotional and physical sensations. Examining the emotional dimension means being open to features of emotional expression, including adopting techniques that reveal and capture this dimension. This is where arts-based approaches and the artist's training are particularly relevant. For example, Saratsi et al. (2019) observes that recent decades have seen research approaches that use the arts, in the broadest sense, in exploring, understanding and representing human actions and experiences, with an increasing number of creative practitioners responding to issues related to environmental change.

> Artists and arts research can offer new perspectives and contribute to transformative ways of exploring, challenging and understanding existing meanings and values as well as creating new ones through the co-production of new knowledge and experiences . . . artistic and creative practices can offer outputs and methods for complementing scientific and environmental understandings, re-envisioning environmental relationships, and improving health and wellbeing, whilst enabling community engagement.
>
> *(Saratsi et al., 2019)*

Arts-based approaches and associated investigation techniques, which take a qualitative approach, provide appropriate tools for revealing the deep meaning behind the landscape, but also act as a means for sensitising the individual to the qualities of the experience (Millman, 2012; 2013; Costa & Coles, 2018; Coles et al., 2013; Coles, 2014; Saratsi et al., 2019; Klein, 2010; Barone & Eisner, 2012).

DOI: 10.4324/9781003099758-6

It is these types of approaches that endorse such enquiry to reveal how the experience can become an embedded one as a combined sensory, physical and emotional experience to give an insight into how landscape encounters are negotiated at a personal level and in a range of circumstances.

In this context, we can also examine the idea of 'geo-emotionality' regarding an emotional, biophilic reaction to the landscape, a concept developed by Dunn (2017). Dunn examined contemporary attitudes to landscape encounters where investigation was focused by his identity as an artist and which, as the term indicates, embraced an emotionally sensitive response to the landscape, articulated within arts practice and explored through various walking and related workshop activities. These concerned developing a heightened awareness of landscape qualities, but also the power of the landscape as a natural mediator and means for exploring historical events and related social issues, how the landscape is used to frame personal, family and other aspects (Coles et al., 2013).

Essentially, Dunn worked as an 'artist in residence' with the remit of exploring contemporary attitudes to a distinctive landscape complex, taking an 'ecocentric' approach involving a focus on ecological associations as basic to enquiry (see, for example Gladwin, 2013 in relation to landscape photography) and which included the associated ideas concerning 'eco-haptics', the sense of touch in relation to ecological associations (www.interhaptics.com/explore/what-is-haptics). These relate to biophilia in emphasising the qualities of the natural environment as experienced by people and how they move through or linger in it (Dunn, 2017; Costa & Coles, 2018). Dunn was based in the Wye Valley, situated in the UK (www.wyevalleyaonb.org.uk/), a location that he describes as an '*imaginatively vital and stimulating landscape*' (Dunn, 2017) and which has particular historical significance regarding the birth of tourism and the development of landscape aesthetics (Gilpin, 2014). In terms of such, it is a location visited by a wide range of tourists, the area being managed by a team that aims to maintain its integrity as an important restorative and ecological natural resource with high conservation and historical status.

Accordingly, such studies confirm the special place of arts-based approaches in ensuring that individuals have the necessary personal tools, awareness, confidence and opportunities to access the positive restorative aspects of nature as part of a life strategy and in:

- Cultivating a 'mindful approach' to the experience of the landscape which serves to stimulate or recharge the biophilic self, an approach which is indicated in the concept of a geo-emotional response, but found in a range of studies.
- Stresses the importance of protecting and providing wild spaces that are freely accessible to the public which are capable of providing a range of ecocentric types of engagement and haptic associations that provide a deep level of sensory involvement via a high quality biophilic experience, typically through simple walking and viewing activities.

The Concept of Geo-Emotionality

Dunn defines his concept of geo-emotionality as the:

> 'embodied affiliation' with a materially specific environment; more precisely, the term is intended to describe corporeally felt emotional sensitivities and to consider apparent physical traces of such encounters'.
>
> *(Dunn, 2017, p. 16).*

Ideas were developed following a period of residency in Japan where his interpretation and understanding of an unfamiliar landscape presented a challenge which was resolved by artistic enquiry, including making, which led to the development of the concept. Geo-emotionality, as discussed here, concerns how an individual responds when encountering the natural environment and how the encounter is 'felt' physically and emotionally, but also how a response may be triggered, captured, represented, then focused and explored through artistic processes, in ways that explicitly reference individual identities and responses.

The concept of geo-emotionality can be considered alongside other theoretical perspectives and ideas discussed in earlier chapters regarding biophilia and landscape perception, affordances and the restorative properties of accessing nature (Chapters 1 and 2) focusing on the quality and emotional content of the encounter as well as moving to initiate a geo-emotional response in others. Such approaches, i.e. engaging with artistic practice and adopting an ecocentric stance based on walking experience, involves a high state of awareness of the physicality of the encounter, the multi-stimulation of the senses and heightened awareness of emotional consequences induced by the varying scene, rhythms of walking and flows of embodied information (Millman, 2013; Costa & Coles, 2018: Coles et al., 2013; Pink et al., 2010; Bannon, 2013).

Art-Based Approaches

Art practitioner-based approaches have a particular relevance when it comes to considering the human dimension and how the environment is perceived, understood and appreciated where:

> environmental artists seek to investigate our human relationship with the environment through embedding their artistic practice within it.
>
> *(www.theartstory.org/, environmental art)*

Approaches embrace the use of reflexive practice as a vehicle and mechanism for acquiring and synthesising information, including modes of representation, enquiry and critique, as well as utilising different communication platforms. For example, Fortnum (2005) suggests that artists possess 'visual intelligence' and that 'artists occupy a position within society, from which they are expected to communicate. . . . Artists perform many roles, one of which is interpreter' (Dunn, 2017, p. 309).

Thus, within this context, artists' interventions present an important suite of tools, activities, props or enabling mechanisms that sensitise the individual, including themselves, to the experience, making it accessible, relevant and meaningful – 'human sensitivity in the relationship of the body in material space' (Dunn, 2017). As such, they are capable of stirring the emotions and gaining insights into the landscape which may lie undiscovered or unrealised, initiating or making evident a bodily felt (embodied) emotional response. This concerns how the body is naturally sensitive to the natural scene, but may need initial priming, and clearly relates to ideas concerning biophilic engagement and biophilia as an innate embodied force.

The geo-emotional approach (Dunn, 2017) thus enquires about the processes involved in accessing and experiencing the natural landscape, using reflexive practice as a vehicle and mechanism for acquiring and synthesising information, including modes of representation, visualisation and enquiry and, as such, uses different communication platforms. These essentially embrace the identity and experience of an artist as the vehicle of enquiry, working alone or with partners, in facilitating or providing the confidence, permission and instruction for others to do the same, enabling them to act in their own 'freelance' capacity relevant to their own circumstances.

> Artists can interpret the visible and invisible and be open to a more extensive and sensory realm than just sight' (Dunn, 2017). . . . Artists can be employed to reinterpret the visible and invisible . . . and refresh stakeholders' overfamiliarity with their surroundings.
>
> *(Dunn, 2017, p. 314)*

Arts and the Environment: Their Influence on Human Behaviour and Attitudes

The arts and environment, of course, covers a very wide range of situations and circumstances (for example, see The Arts Story, *www.theartstory.org/*) which includes how the arts can be used to promote pro-environmental behaviours or challenge existing norms. Despite this, Curtis et al. (2014) make the observation that 'the environmental and social psychology literature is largely silent on the role of the arts in affecting environmental behaviour' – yet it is clear that arts-based methodologies are particularly effective in revealing the quality of an environmental encounter in ways that are defined by the individual, including the relevance of the experience in respect of life strategies (Millman, 2012; Coles et al., 2013; Costa & Coles, 2018).

Curtis et al. (2014) consider the role of the arts within the broad theme of 'working towards ecological sustainability', investigating a range of case studies and how they influence environmental behaviour and attitudes under the twin themes of 'environmental sociology' and the 'sociology of the arts', where the arts are seen as an important tool regarding environmental education (Alexander, 2003).

FIGURE 5.1 Environmentally themed mural: an appreciation of the aesthetic and intrinsic value of nature can be catalysed by artistic interventions, to sensitise people to the properties and importance of the natural world.

Curtis et al. reviewed a range of environmental arts projects and interventions – for example, held in conjunction with ecological restoration projects – and considered how they form pathways for the communication of eco-based information, in creating empathy for the natural environment and how the arts, e.g. arts installations, are included in ecologically sustainable development to reinforce and celebrate messages or achievements (see Box 5.1; Figures 5.1 & 5.2).

FIGURE 5.2 Unlocking the River Severn: fish sculpture celebrating the return of the Twaite Shad to the river, shown here in its temporary location.

BOX 5.1 THE INFLUENCE OF THE ARTS REGARDING CHANGING BEHAVIOURS AND REINFORCING ASSOCIATED ATTITUDES TO THE NATURAL ENVIRONMENT

- **Values, beliefs and attitudes** – the arts are considered important in determining environmental behaviour, where an appreciation of the aesthetic and intrinsic value of nature can be catalysed by interventions.
- **Awareness of consequences** – through an educative function, including improving knowledge, where the arts are a potent means of conveying ideas.
- **Unfreezing habits** – by embracing the celebratory aspects of the visual and performing arts in relation to eco-based work, thus removing barriers to adopting or understanding pro-environment behaviours, e.g. through ecological restoration rituals where arts practice can supply the tools.
- **Social norms** – affirming pro-environmental attitudes and behaviour by signalling social approval and helping to build human capacity in managing the environment more sustainably, fostering collaboration.
- **Self-concept** – influence the construction of self-identity, e.g. by exposure to eco-based festivals and performances.
- **Physical structures** – through the establishment of structures via community and public art incorporated into urban planning that further endorse an ecocentric stance.

Based on Curtis et al. (2014). Towards ecological sustainability: observations on the role of the arts, S.A.P.I.EN.S [Online]. (7)1. Available at: https://journals.openedition.org/sapiens/1655, [Accessed December 2021].

Notes: 1. The main categories are those identified by Curtis et al, although slightly re-organised.
2. The explanatory notes are modified from Curtis et al.
3. Curtis et al. developed their categories from reviewing a range of environmental art projects from which they were able to identify a broad set of pathways. For more information and original wording, the reader is referred to the original source.

The Eco-Haptic and Experiencing the Landscape

As introduced earlier, the 'eco-haptic' (Dunn, 2017; Gladwin, 2013) is a term adopted in relation to an ecocentric (ecologically focused) approach and understanding of landscape imagery and how 'reading' such imagery actually involves a broad collection of sensory stimuli, reflecting on and how the body moves through the landscape, including being more consciously aware (mindful) of the process of engagement, and how the physical and emotional senses are responding. It recognises that the environment, although often recorded as an image, is actually perceived as a deeper embodied and multisensory experience which has specific personal relevance and meaning. Eco-haptic readings of the landscape thus involve an ecocentric approach, and relate to how the body moves through the environment and is stimulated by different sensory cues, i.e. the situation is embodied and involves an empathic view of the world (Dunn, 2017).

As an artist, Dunn (see Box 5.2) explored the idea of the eco-haptic and geo-emotionality, but it is also easy to see the same or similar themes being expressed within other theories of landscape perception discussed in earlier chapters. These also emphasise that the ways in which we read and associate with the landscape are constructed on a personal level, including via dialogue with eco-based, biophilic parameters (Choudhry, 2015; Costa & Coles, 2018; Coles et al., 2013). These involve a high degree of emotional attachment, which is realised through multi-sensory or bodily engagement (Pink et al., 2010), i.e. an empathic and embodied response related to self. Ecocentric, eco-haptic and geo-emotional ideas are thus clearly indicated in the concept of biophilia, but the endeavours of arts-based practitioners, who readily explore the emotional dimension and encourage others to do the same, appear to have particular agency in moving people towards a deeper understanding and appreciation of landscapes in relation to self. In this respect, such initiatives can be understood as a supplementary pathway to achieving biophilic fulfilment through deep environmental engagement.

BOX 5.2 TOUCHING TOPOGRAPHY, NEGOTIATING LANDSCAPE ENCOUNTERS (DUNN, 2017)

- Dunn investigated human encounters with the landscape, taking on the mantra of a tourist, where his artistic processes included the recording of personal encounters using a digital camera or smartphone camera as well as an audio recorder and the writing of notes. The process involved one of immersion in the landscape, moving through it and reflecting on the experience.
- Dunn studied the environment by familiarising himself with the physical aspects of the landscape, allowing different bodily felt sensations to influence him emotionally as he moved through it, being aware of the 'embodied affiliation' and 'corporeally felt emotional sensitivities', reflecting upon the experience by maintaining a creative diary.
- Methodology included different ways of sharing ideas and publicising activities as a contemporary practitioner, including using an online blog and embracing the qualities of the internet as a tool to critique, explore and share information.
- Research was initially based upon immersion in the landscape in the manner of an artistic residency, with the notion of getting closer to nature/the landscape by dwelling in it and being in close contact with it, but it involved social isolation and an unnatural removal from the normality of social and family engagement. Accordingly, guided by the initial stages of the study, follow-on work involved walking with others, which gave more emphasis to social exchange set within the environmental encounter, considering how others explored, experienced and expressed their own geo-emotional interactions and how these were mediated by the features of the walk, rhythm of pace, scene and cultural context.
- As the research progressed, it appeared to entail a move from experiencing the landscape as an outsider, i.e. tourist, to an insider involving dialogue with others, including the AONB management team responsible for the area, as well as activities as such as group walking and targeted theme-based explorations, culminating in encouraging other groups to develop their own ideas and expression to reveal the connections and flows associated with the negotiation of landscape encounters.
- Different strands of the research served to further validate the idea of geo-emotionality, considering how a range of artists residing in the location had been specifically influenced by the landscape and sought to dwell in it, both metaphorically and literally, and how this was expressed in their work, varying from photography to painting and ceramics; this gave further insights into the ways that biophilic connections and geo-emotional experiences were embraced, represented and communicated through works of art.

The Nature of the Landscape Experience and Arts-Based Processes of Enquiry

By adopting the perspective and techniques inherent in artistic enquiry and by embracing an ecocentric approach, it is possible to explore how the experience manifests itself but also how it can be supported in others. For example, by setting up a residence in the heart of the landscape complex, Dunn established a situation where the different aspects/dimensions of the landscape were experienced on an everyday basis, through the seasons, weather patterns, observing and recording, immersing himself in the experience – 'intended to harness the proximal materiality of the landscape as an emotionally provocative medium' (Dunn, 2017, p. 227).

> Whilst moving through the landscape, I periodically stopped and typed a short entry into my mobile phone. I intentionally used a poetic structure, utilising alliteration and rhyme for example, as I have previously done before in my artistic practice. I used descriptive language that reflected my movement through space as a way of mirroring what I was seeing.
>
> *(Dunn, p. 137)*

> Moss flocked roots, finger like flow below . . . bright broccoli topped woodlands invite my hands to caress depicts the way in which I wanted to handle the far off forest that I could see, to bring it closer to my grasp.
>
> *(Dunn, 2017, p. 137)*

The language used by Dunn is sensual and poetic, conveying an immersive situation and the 'draw' of the landscape where Dunn's 'mirroring' promotes an alignment of body, self and environment, provoking an emotional response but which also initiated a desire or need for more intimate association, to explore what was seen through other senses, especially touch: 'bring it closer to my grasp'. This was further explored through diary entries, notes and drawings, but also by the practice of blogging (see Merry, 2010; Pritchard, 2008), a process which formed a routine part of his professionally related arts practice, i.e. using an online blog and embracing the qualities of the internet as a tool to critique, explore and share information. Dunn thus applied his routine practices to instil a depth of awareness of the properties (the biophilic attributes) of his landscape experiences through 'word images' and sharing to rehearse and re-affirm his encounters and define their power in relation to his own identity.

Accordingly, this (and subsequent group walks) applied ideas where the findings allow further consideration of the connections and flows inherent in biophilic associations. As Dunn states,

> the landscape as an entity experienced through real and imagined touch . . . it was my intention to encourage people to use their hands . . . to express the experience in place as opposed to using purely visual means of recording, such as a camera.
>
> *(p. 227)*

Via group walking, Dunn encouraged a mixed group of university students

> to use mobile phones, as a contemporary image gathering technology, to obtain visual image from the location . . . To further clarify my understanding of the images produced by the students, I asked each to submit accompanying text, to ensure that their aesthetic intentions were fully understood.
> *(p. 234)*

Observations were made regarding how participants recorded the event, how they reflected on the experience, inviting personal feedback and requesting comments associated with selected recorded views. At the same time, exposure to any pre-conceived anticipations of the area was avoided, since such might influence expectations (for example, published visitor reviews, photographs or related statements) (Dunn, 2017).

Walks intentionally included a range of interactions, experiencing the enclosed wooded (occluded) landscape with limited sight lines, as well as the opposite situation, involving the revealing of views, opportunities to stop in different environments, including varying intensities of physicality, topography, ground conditions and opportunities to touch vegetation or record events, including to observe and, if minded, photograph close or distant views and thus explore the type of situations that fostered different emotional, visual and haptic connections to the landscape.

- *Moving through the occluded landscape* – it was clear that the effort involved in negotiating the steep topography involved different amounts of exertion, pace and footfall, as well as restricted sight-lines, which together with goal-obsessed behaviour to reach the top (Dunn, 2017) led to a certain mode of walking and type of experience, with particular focus on the physicality of the task, negotiating the landform and concentrating on footfall. This changed when the summit of the wooded hill was reached, but appeared to have sensitised participants to the ecological conditions: 'we were more aware of our surroundings after the steep incline, with a focus of making it to the top' (group student walk participant, Dunn, 2017, p. 253).
- *Emotional intensity and foraging behaviour* – In contrast, moving from woodland to open meadow initiated a profound and sudden emotional release, a collective 'wow' (Dunn, 2017) as extensive views were suddenly revealed. This prompted a different type of behaviour involving exploring the qualities of the landscape, the taking of many photographs, including 'posing in the scene', resting in the landscape, touching and further exploring the properties of the location. For example, for one participant it included a spiritual dimension when he unravelled a prayer mat, expressing his desire to say prayers 'in a beautiful place'. (Dunn, 2017, walk participant). For others, the experience was clearly embodied in childhood and family memories triggered by the specifics of the encounter, touching the tall grass and observing many butterflies. The discovery of orchids caused further excitement and exploration as 'something special', accompanied by further photography and searching among the vegetation an action which can be expressed as a (contemporary) 'foraging activity'.

Interpreting the Embodied Biophilic Experience

In considering these ideas, it is possible to make some connections concerning experiences of biophilic situations and the way that they are recorded on a personal basis. This includes the concept of affordances and perceptual theories about viewing the scene (see earlier chapters) and follow-on behaviours that involve exploration of 'habitat' as an emotionally and physically satisfying activity providing appropriate sensory feedback in terms of biophilic needs (sensory affordances). In doing so, several ideas can be considered:

- Viewing the scene and taking an image of the scene, e.g. using a mobile phone, relate to an initial and easy assessment of the affordances or potential of the site for human activity. In the context of biophilia, these relate to innate biophilic needs, but are now realised in the reality of contemporary recreation and relaxation.
- The routine use/availability of the mobile phone and digital camera and other mobile devices now make photography and video a default position for observing the landscape. Thus, in the context of ecocentric understanding, can be considered as an integral part of observation and interpretation.
- It seems likely that contemporary communications platforms are capable of forming a facilitator in realising and embedding the biophilic experience, and thus possess 'agency' in in this respect; these might be encouraged as self- or community-applied tools involving the posting of images and related material on web-based social media and blogging sites.
- In the context of recreation and leisure, exploration activities regarding the potential of the environment to satisfy biophilic need can be considered as a type of 'contemporary foraging' activity – 'visual and sensory foraging' – exploring, testing, touching, smelling, observing, watching, collecting and looking for, but set within the context of recreation rather than basic survival.
- These actions involve activities which provide high-quality feedback concerning haptic connections and associated emotional satisfaction, a search for visual and sensory excitement and contentment fulfilling biophilic need. Associated web-based follow-on activities can be interpreted as part of this activity as experiences are explored then relayed and responded to by others.
- Although these activities are now divorced from basic survival, falling within a recreation context, they are still sought out as part of a life strategy.
- In this respect it is possible to suggest that biophilic need is satisfied through 'recreational foraging', embracing the associated eco-haptics and other senses – walk through, touch, smell, pick up, splash through, record and communicate. It is noted that, in the respect of eco-haptic properties, many urban landscapes can be considered as impoverished or generally not available for such activities.
- Eco-haptic connections are also forged in negotiating the terrain, where more rugged conditions require careful observation and more complex navigation, where the body responds to the physicality of the task and notices such aspects as footfall and exertion.

- In the absence of restricted routine involvement with the natural world, i.e. as indicated in respect of urban living where nature is under-represented, the full cycle or depth of activities are not routinely experienced while, in addition, people may not have the confidence or skills to experience them.
- People may benefit from some kind of support in realising the full impact of environmental qualities including initiatives that involve arts practice, artists working in the community, co-collaboration ventures or web-based platforms.

Arts-Based Actions and Developing a Mindful Approach

Because of their unique properties, arts-based approaches are being increasingly explored in relation to environmental issues. For example, Curtis et al. (2014) observes that 'it is likely that the arts sector will be an important structural component of ecologically sustainable societies in the future'. In addition, Fien (2003, cited by Curtis et al.) suggests that appreciating the aesthetic and intrinsic value of nature through art is important to people's willingness to conserve it and provides avenues for pro-environmental interaction. Furthermore, such approaches clearly support the realising of personal biophilic contentment, enabling nature to feature more consciously as a positive emotional force, catalysing the adoption of environmentally supportive behaviours, including improving knowledge and increasing awareness of our own actions and sensibilities.

> Like biophilic design integrates nature into our cities, biophilic art is a process that helps us to presence a connection with our natural selves. By summoning up the senses and sense experience, and becoming present in and with nature, a pause moment is created, which develops into a more discerning emotional awareness of our wellbeing and that of the ecological world we impact.
>
> *(www.biophiliccities.org/biophilic-art)*

Many artists use their work to communicate important insights into human relationships with the environment (Pollak & MacNabb, 2000; Collins, 2008; Mills & Brown, 2004), while citizen groups have promoted nature conservation and environmental protection through community arts projects (Mills & Brown, 2004), including festival events to celebrate environmental excellence, stimulate awareness, responsibility and underpin identity. One such example is the River Wye Festival in the UK, which embraces drama and fantasy to raise awareness of the environment and natural heritage of the area (https://wyevalleyriverfest.com). The festival involves a partnership between theatre groups, the Wye Valley AONB and the Arts Council, as well as local sponsors and community champions using art to both celebrate the River Wye as a wonder of nature, but also as a means of engaging the community and sensitise them to the properties and importance of the natural world and local features.

FIGURE 5.3 Reflection of self in the landscape. Using an art installation to heighten awareness of the body located in the landscape.

Led by artists and communities creating work with the environment at its heart. We bring local people, environmentalists, and talented artists together in an internationally important landscape, for a creative exploration of what is possible. We support and drive a programme of community engagement and co-creation which culminates in a unique cross-border biennial arts and environment festival: The Wye Valley River Festival.

(https://wyevalleyriverfest.com/)

On a more individual level, artists work by placing installations in the community, specifically targeting urban situations. Several are detailed by Curtis and other sites (see www.biophiliccities.org/biophilic-art); these include Jamie Jackson's projection of environmental images onto walls (www.jaimejackson.org/) and Carmen Bouyer's film 'Seeds', filmed in New York's urban forest Inwood Hill – 'a poetic attend to grasp ways in which one can expand the experience of the self to embrace its wild environment'. Other ideas include initiatives that are designed to encourage deeper experiences of the environment and provide instructional, web-based material. For example, 'Listening to the Sounds of Silence', hosted by Childhood by Nature (www.childhoodbynature.com/), advocates a deep listening and sensory activity to encourage a mindful approach.

> Sensory information that can help restore your child's mind and help develop their body with the many smells, textures, temperatures and colors . . . when we tune in, we hear sounds all around us – a cacophony or orchestra of the sounds of nature, revealing the life that is all around is, some hidden and some distant. This quick 15-minute 'sound of nature' activity will help your child discover a whole new world, be more mindful and more tuned in to nature.
>
> *(childbynature.com)*

Guidance suggests a quick 15-minute 'sound of nature' activity in a quiet spot away from traffic, giving instructions to parents which involve exploring the sound further, being aware of sounds, focusing down, describing the sensations, identifying the source and, finally, using the web-based app 'ChirpoMatic' to identify any birds involved (www.chirpomatic.com). As well as encouraging a deep approach, essentially this also conforms to a foraging (explorative) activity; similar types of initiatives are advocated as part of the Forest School approach, involving elements of discovering, touching, collecting and making, which encourages deeper engagement within the context of play (see Forest Schools and the outdoor classroom studies, Chapters 6 and 9).

Web-based activities have also been explored in relation to the environment, where Merry (2010) analysed the personalised content of related blogs and considered that the potential of blogging to influence grassroots support for environmental groups could be greatly extended to inform local communities. However, it was also observed that, while discussion channels provide effective participative online communications, they tend towards issues such as climate change, rather than as a supplementary suite of tools to support general, local or personal environmental awareness. However, electronic platforms are used effectively to keep residents/communities up to date in the form of newsletters and events such as nature observations, planned walks, conservation activities, conservation guidance and other activities and form a highly accessible communication platform; it seems quite likely that their use will become increasingly sophisticated in providing simple tools and techniques that support the development of a refined and satisfying biophilic experience (see Chapter 9).

In this connection, we can also highlight the ideas such as self-talk and self-narrated walking applied by Millman (2012) and Costa (Costa & Coles, 2018) as an investigative tool (Chapter 4), but which also supported a deeper engagement with the environment and the acquisition of a mindful approach which seems to persist even after the original encounter. In the technique, it is easy to use mobile devices to record thoughts as one moves through the landscape, stimulating an awareness and associated pleasure in the experience and moving into a more intense state of emotion which appears to transcend space and time, typical of that featured in a positive flow experience (Csikszentmihalyi, 1990). The technique has been applied in both natural and urban situations and in different urban communities, with equally effective results and positive feedback from participants.

Conclusions

This chapter has explored the emotional dimension of biophilia. It considers the role of arts-based practice, and that the arts have agency in supporting people to appreciate and experience the intricate qualities of the natural environment and related biophilic places, thus to move to a deeper level of engagement. As such, it is clear that arts-based techniques and collaboration provide a complementary pathway to achieving effective deep biophilic interaction generally associated with adopting a mindful approach.

The synergies between arts and the environment are briefly discussed, where the arts are considered important in determining and supporting pro-environmental behaviours at a range of levels, including deeper appreciation of the aesthetic and intrinsic value of nature, the construction of self-identity, awakening the biophilic self, fostering environmental awareness, including the consequences of actions, unfreezing habits and challenging accepted norms.

Discussion introduced the concept of geo-emotionality featuring a study of the Wye Valley, a recognised high-quality landscape, which was explored through the activities of the arts practitioner, specifically examining the perspectives gained working in and processes involved in moving to a deeper level of engagement involving experiences which have particular emotional intensity. Ideas are situated within landscape perception theories, but the process of deep engagement involving the eco-haptic (and other sensory modes) is conceptualised as a contemporary form of recreational foraging, whereby the fine qualities of the environment are explored in terms of their properties, and which form biophilically and emotionally satisfying experiences.

Discussion points towards a situation where people would benefit from support from the arts sector alongside other mechanisms in developing the confidence to realise a deeper positive embodied biophilic experience and acquire an associated mindful approach. This is a situation increasingly addressed by the arts disciplines in exploring arts, the environment and nature, as well as working across disciplines. Several examples of approaches are discussed to illustrate the value of arts/environment-based approaches.

In suggesting ideas, particular emphasis is placed on the provision of accessible, stimulating, nature-based, biophilic-focused landscapes, close to and within urban conglomerations. It would be in these 'quality' places that people can realise a contemporary form of visual and sensory foraging as an activity set in a modern recreation context, gauged by the extent by which they embrace haptic and emotional connections and sensory affordances; however, it is noted that many urban situations do not afford such opportunities.

References

Alexander, V. D. (2003). *Sociology of the Arts: Exploring Fine and Popular Forms.* Malden: Blackwell Publishing.

Bannon, F. (2013). Synchronising self and city. In Coles, R. W., Millman, Z. (eds). *Landscape Well-being and Environment.* Abingdon: Routledge, pp. 38–52.

Barone, T., Eisner, E. W. (2012). *Arts Base Research.* Thousands Oaks: SAGE Publications Ltd.

Choudhry, K. Z. (2015). *A Multimodal Measurement Approach Using Narratives and Eye tracking to Investigate Visual Behaviour in Perceiving Naturalistic and Urban Environments,* (PhD Thesis). Birmingham: Birmingham City University.

Coles, R. W. (2014). Environmental interaction and engagement – Supporting well-being. In Burton, E., Davies-Cooper, R., Cooper, C. (eds). *Wellbeing and Environment,* Chapter 17. Oxford: Wiley-Blackwell, pp. 445–499.

Coles, R. W., Millman, Z., Flannigan, J. (2013). Urban landscapes – Everyday environmental encounters, their meaning and importance for the individual, *Urban Ecosystems* 16(4), 819–839. https://doi.org/10.1007/S11252-013-0327-Y

Collins, T. (2008). Lyrical expression, critical engagement, transformative action: An introduction to art and the environment, *Ecologie Politique* 36(2), 127–153. https://doi.org/10.3917/ECOPO.036.0127

Costa, S., Coles, R. (2018). The self-narrated walk. A user-led method to research people's experiences in urban landscapes, *Landscape Research* 44(5), 526–540. https://doi.org/10.1080/01426397.2018.1467004

Csikszentmihalyi, M. (1990). Flow: The psychology of optimal experience, *Journal of Leisure Research* 24(1), 93–94.

Curtis, D. J., Reid, N., Reeve, I. (2014). Towards ecological sustainability: Observations on the role of the arts, *SAPIENS* 7(1). http://journals.open edition.org/sapiens/165D.

Dunn, G. P. (2017). *Touching Topography: Negotiating Landscape Encounters with Several Parts of the Wye Valley Area of Outstanding Natural Beauty,* (Doctoral Thesis). Birmingham: Birmingham City University.

Fien, J. (2003). Learning to care: Education and compassion on JSTOR, *Australian Journal of Environmental Education* 19, 1–13. www.jstor.org/stable/44656360

Fortnum, R. (2005). *What Is Visual Intelligence and How Do Artists Use It? Visual Intelligences Research Project.* London: Lancaster University. www.visualintelligences.com/visual-intelligence-how-artists-use-it.html

Gilpin, W. (2014). *An Essay upon Prints: Containing Remarks upon the Principles of Picturesque Beauty.* Cambridge: Cambridge University Press.

Gladwin, D. (2013). Eco-haptic photography: Visualising Irish Bogland in Rachel Giese's the Donegal pictures, *Photography & Culture* 62(2), 157–174. https://doi org/10.2752/175145213X13606838923237

Klein, J. (2010). What is artistic research? *Gegenworte* 23, 25–28. www.jar-online.net/what-artistic-research

Merry, M. K. (2010). Blogging and environmental advocacy: A new way to engage the public? *Review of Policy Research* 27(5), 641–656. https://doi.org/10.1111/J.1541-1338.2010.00463.X

Millman, Z. K. (2012). *Landscape Narratives and The Construction of Meaning in The Contemporary Urban Canal-scape*, (Doctoral thesis). Birmingham: Birmingham City University.

Millman, Z. K. (2013). Photographic postcards as research tools: The postcards from the cut study, *Graduate Journal of Social Science* 54–75.

Mills, D., Brown, P. (2004). *Art and Wellbeing*. Strawberry Hills: Australia Council for the Arts.

Pink, S., Hubbard, P., O'Neill, M., Radley, A. (2010). Walking across disciplines: From ethnography to arts practice, *Visual Studies* 25(1), 1–7. https://doi.org/10.1080/14725861003606670

Pollak, M., MacNabb (2000). *Hearts and Minds: Creative Australians and the Environment*. Alexandra: Hale and Iremonger.

Pritchard, D. (2008). Artistic licence: A review of the forestry commission's involvements in the arts, and options for the future, *SSRN Electronic Journal*. https://doi.org/10.2139/SSRN.2592367

Saratsi. E., Acott, T., Allinson, E., Edwards, D., Freemantle, C., Fishe, R. (2019). Valuing nature, valuing arts & arts research, *Valuing Nature Paper VNP22*. https://valuing-nature.net/valuing-arts-and-arts-research

6

LISTENING INTO NATURE

Exploring Children's Biophilic Experiences of Urban Wild-Spaces

Sharon Watson and Richard Coles

Introduction

Writing in 1978, Colin Ward published his acclaimed book *The Child in the City* (Ward, 1978; see also Rissotto & Giuliani, 2006), which was, as Ward states 'an attempt to explore the relationship between children and their urban environment'. Ward examined how children engaged with their urban worlds, what they did, how they used different spaces, but he also suggested 'that something has been lost in this relationship', speculating about the ways in which 'the link between city and child can be made more fruitful and enjoyable'. Ward looked at the totality of the city and how children used the diverse qualities of the environment in their play activities. Moving forward, similar issues are still being raised, but in this case, specifically in relation to the urban child's exposure to nature, in particular the opportunity to access and experience nature on their own terms. One of the main commentators is the author Richard Louv; Louv suggested that children are experiencing 'Nature Deficit Disorder' (Louv, 2005) and are suffering as the opportunities to access nature and experience the properties of the natural world have diminished. Like Ward's observations, the concern is that something has been lost.

Following through these assertions, there are indications that lack of exposure to nature, perhaps better put as ensuring an essential exposure to nature, can be addressed by understanding the biophilic properties of urban nature locations, not as defined by adult categorisation, but by the child's experience. Children clearly value the natural experience (Mustapa et al., 2015) and a range of authors emphasise that early exposure to nature creates an enduring relationship which is carried through to adulthood and thence to future generations (for example, see Chawla, 2007; Derr & Rigolon, 2015).

As discussed in earlier chapters, the indications are that, especially within the context of city life and isolation from the natural world, the substantial benefit

DOI: 10.4324/9781003099758-7

accrued from biophilic contact is crucial to well-being and will most probably have their origins in childhood experience, but that its cultivation needs to be stimulated just like any other aspect of childhood development. In fact, it can be argued that failure to provide appropriate opportunities for such introduces a penalty regarding future opportunities to access the restorative properties that can be accrued. The linkage of biophilic experiences to well-being and developing the confidence to access such experiences would seem critical to childhood development (see for example, White & Stoecklin, 2008).

The follow-on from this argument is the question: how do young and specifically urban children experience or explore the natural environment and related wild-spaces, and can we understand the processes involved?

Children's Relationship with the Environment: the Context of Play and Play Affordances

A range of authors comment on the playful encounters of children. As already identified, these include Colin Ward's *The Child in the City* (1978), and more recently Lester's (2011) study which considers children's lives and how they appropriate spaces and enjoy appropriating spaces for their own social needs – how (urban) space is utilised. Lester emphasises that: 'Playing is opportunistic, appearing in the cracks and crevices which are always present in space, no matter how far powerful structures seek to close down possibilities for action'. Play is ubiquitous, it happens everywhere where an opportunity presents itself, while Moore discusses how children like reworking places (building, dens, slides, etc.) and how using their imagination is important, rather nicely put as 'being in control of out of control' (Moore, 1987 p. 51). Other authors (Lester, 2011; Kallio, 2009) emphasise that children are good at using embodied means to realise their play objectives, and this is how the embodied nature of playing is demonstrated in their relationship with the environment, which is particularly interesting in the context of biophilia and nature. Essentially these authors talk about the 'play affordances' offered by places, how children approach and interact with their outside environment through embodied responses, the indicated physical actions and sensations.

Children respond to environmental cues and invitations in nature to do something and that through the embodied aspect of biophilia are particularly 'excited' by these natural cues as opportunities for positive actions are recognised and realised; however, as these authors identify, play is ubiquitous and opportunities are realised in what adults might consider the least appropriate situations (Moore, 1987). Child-related affordances are specifically defined by children, not adults; moreover, it seems likely that, when attempting to identify what they are, it is important to engage within the remit of the play situation rather than rely on adult-based criteria. Play, then, is the child's natural means of exploration and environmental literacy develops in children as they experience and read the environment in terms of its key elements and associated opportunities with the potential for embedding powerful and meaningful memories.

Two examples drawn from different disciplinary contexts perhaps help to illustrate the point.

- Ergler and Kearns (2013), looked at 'children as explorers', examining how they use their local urban environments (in the context of New Zealand); they cite the example of Josh's experience of the 'bush' environment and relay that 'for the "bush walk", where it was anticipated that nature played a favourite aspect of the local environment [Spencer & Blades, 2006] it was the personal relationships and fun that was memorable'. Specifically reporting from Josh's narrative, Josh revealed that 'the time in the bush was his favourite aspect of the walk, but not because of the natural environment or the views, which were highlighted in their talk about the photos, rather the landscape was the arena in which a fun activity took place, but spoken word and personal interaction created the happy feeling . . . the boys clearly demonstrated that activities and scenery play an important part in their well-being and enjoyment'.
- Marcia Nell & Walter Drew, writing for the NAEYC (The National Association for the Education of Young Children, see Nell & Drew, 2022) and discussing the 'Five Essentials to Meaningful Play', start their discussion with a quote: 'one of my clearest memories is slithering through the tall grass like a snake with my brother and sister'. In this 'memory' we can see the biophilic interaction and bodily alignment with the environment experienced through play. This is given as an example of 'meaningful play' involving childhood joy, family and emotional bonding, which persists even long after children are grown, i.e. into adulthood. This type of memory and the obvious intensity of the experience is clearly the result of a biophilic connection and bodily engagement, involving a specific alignment with the ground, movement, a joint playful experience and an element of fantasy existence and exploration which is embedded as a positive and treasured memory.

Josh's 'bush walk' and Nell & Drew's 'slithering through the grass' both serve to illustrate the quality of the child's biophilic experience in the context of play and also sets the context for further discussion regarding how the biophilic interface is demonstrated when groups of city children access natural and wild places located within their neighbourhoods.

It appears that children have a strong biophilic response, i.e. as the biophilic hypothesis suggests, their biophilic self is contained as an innate property, and the nature connections made in early life readily act as a conduit for later life experience, which can be seen across and between generations and situations where mental restoration is sought (Prisk & Unsworth, 2018; Gill, 2014; Mustapa et al., 2015; Louv, 2005; Thomas & Thompson, 2004). This concerns the biophilic experiences of urban children, city children that dwell in the modern urban world and issues concerning how they interface with nature, the nature of their environmental experience and the various facets of this relationship (Yanez et al., 2017).

Sense Data, the 'Biophilic Self' and Experiencing the Outdoors

Looking at the meaning that children derive from environmental experience involves how the individual gains or constructs meaning from the encounter, an approach that draws on 'sense data' and mental processes involved to define 'self'. In relation to nature and ideas of biophilia, essentially, this means an investigation of the 'biophilic self'. Again, according to the theory of biophilia, biophilia would be expected to be present in children as an innate property, although it may need awakening and stimulation as a latent positive force. Accordingly, this moves us to examine different ways of talking about nature (Watson, 2017; Kroh & Gimblett, 2007) and how concepts of nature are located within 'embodied experiential processes' (see Watson, 2017 and also Bunkse, 2007; Lund, 2012; Thwaites & Simkins, 2007; MacPherson, 2010).

Studies on children's lives recognise that moving outdoors is a qualitatively different experience for the body: 'Outside, participant movements, gestures and embodied/bodily engagement create a particular category of sense data, one that is occurring in-the-moment and subject to ephemeral, shifting encounters'. (Watson, 2017; see also Noe, 2004; Gibson, 1986; Thrift, 2000; MacPherson, 2010; Lury, 2010; Kempen, 1996). Accordingly, such ideas concern how children situate themselves in the natural world, but also how the natural world is situated in children or how the biophilic response is situated within childhood experiences. Interest here lies within the synergies between human biology and rhythms of nature and how early experiences contribute to child development and future actions, 'a way of knowing the World that is not replicated through indoor learning' (Watson, 2017, p. 20; see also Cobb, 1977).

Importance is placed on the 'thrill of embodied sensations' (Watson, 2017; see also Coster & Gleeve, 2008) which reflect the haptics of the body in motion and 'are filled with embodied responses to natural phenomena' (Watson, 2017, p. 26).

Discussions regarding nature experience and biophilic interaction thus emphasise 'that physical immersion and proximity to natural settings evokes a different tuned biological self' (Watson, 2017 p. 29) and such studies start to reframe how nature is conceptualised; for example, the flows and fluxes of the weather world (Ingold, 2008), being in the sunshine, wind and rain, i.e. in terms of the embodied experiences of these rhythms and natural cycles and how they are cued in nature through seasonal change, sounds, the rustling of leaves etc. However, especially in relation to modern urban-based living, there is criticism that city spaces and nature spaces have become polarised, which underplays the intermingling of humans and nature in cities (Hinchcliffe, 1999) and in many situations, an insulation from the natural flows and fluxes, especially in relation to everyday childhood opportunities for such experiences.

Various authors discuss the experience of landscape and natural phenomena, including the famous child psychologist Jean Piaget (1973), who gives details of research conversations with children where they believe that they have a causal relationship with natural phenomena, the movement of the clouds, the moon or stars, but which declines as culturally determined ways of thinking and the acquisition of related knowledge alters the perspective of self. The development of fantasy worlds also features, for example, as Johnson (2000) states, 'supermen, ghosts, dinosaurs and dragons', as does 'magical thinking'.

Despite the typical rationalisation of nature's properties through science and cultural beliefs, within the landscape experience, magical thinking does seem to persist in adult personal accounts of nature. Watson (2017) cites several examples:

> 'The sun was saying to me: this is the best place in the world'.
>
> *(Gray, 2003)*

> 'She felt as though it (nature) had seen through her very substance'.
>
> *(Ohta, 2001)*

> 'Soaking in earth bound senses'.
>
> *(Bunkse, 2007)*

Expressions of the properties of the natural and nature given in personal narratives obtained from those when accessing the landscape often contain references to the magical qualities of the experience rather than relying purely on conventionally rational or scientific explanations, e.g. regarding the experience of water, trees and nature's sounds (Brady, 2002). It is probable that these associations are developed in early childhood as a direct response to exposure to nature and natural systems, and which endure into adulthood as part of individual self/world perspectives, i.e. as part of the biophilic self, although they are usually overlain with cultural and scientific values.

The significance of magical thinking about the environment is highlighted by Nemeroff and Rozin (2000) as serving important cognitive, emotional social and adaptive functions which have retained a place in human experience across cultures and through history; as Watson (2017) considers, landscape experiences 'combine physical magical and fictional realities evoking sensations of self which are derived from biophysical engagement/response to natural phenomena and landscape atmospheres and appear important in facilitating connection with nature', and by implication creating a positive biophilic response, while the added use of fantasy makes sense to children when other explanations fail or are inadequate. However, others suggest that magical thinking, while evident in children's biophilic experiences, may be less overt in adult life; in addition, the presence of adults inhibits it in children, and it can remain hidden unless appropriate ways are present to identify it, or indeed it may not be recognised or simply dismissed by adults (Woolley, 1997).

Daydreaming is also highlighted as a central feature of naturally occurring thought (Singer & Singer, 1992) and also indicated in Attention Restoration Theory (Kaplan & Kaplan, 1989), which involves movement into a realm of reverie, a sense of being away, fascination and absorption in the moment, where it is clear that self and environment become aligned through the experience of nature.

BOX 6.1 DAYDREAMING SUPPORTS

1 A sense of identity through imagination.
2 Absorbed thinking, musing and immersion in a temporary alternative realm.
3 A temporary distraction from external cues.
4 Empathy, being good at listening to others.
5 A kind of self-entertainment; helps maintain sanity.
6 May be a long-term effort to cope with issues in life.
7 A sense of fascination.
8 A sense of being away.

Notes: Items 1–6, see Singer and Singer (1992) regarding daydreaming in natural thought; for items 7 and 8, see Kaplan and Kaplan (1989) regarding Attention Restoration Theory.

Eliciting Children's Biophilic Worlds

Children's engagement in nature is driven by sensory cues and biological engagement, but what it means to children and how they express it is a rather more complex question. There are critical issues regarding how adults, researchers and professionals might approach the subject, including the intergenerational responsibility placed on such when working with children, education, interviewing, or interpreting findings, giving the child a voice and accepting its validity. In this context, Watson (2017) emphasises the importance of practitioner-led approaches, and how practitioners use different methods to unlock how children experience and imagine the world, giving examples of approaches involving photography, drawing, conversation, performance, model making, music and dance (Watson, 2017; Ewald, 2000; Reding, 2009; Duckett, 2009). In particular, the emphasis is on child centric approaches, the child as researcher, appropriate involvement and finding 'truths' through a variety of engagement mechanisms.

The way that we respond to places can be explained by the concept of 'affordances' (see Chapter 1) and, in the case of the child, these appear principally related to play opportunities. Places are recognised in terms of their biophysical properties and those which have strong biophilic properties, i.e. they have natural qualities, also constitute high quality play affordances. The experience which is recounted by the child involves the realisation of these affordances involving the kinaesthetic and haptics of movement, moments of intensity, thrill and associated meaning.

Children's interaction with such environments involve what can be called '*imagined trajectories*'. Imagined trajectories relate to the individual as they experience and make sense of the environment. They involve the connections that are made and the imagined placings of things in relation to bodies and landscape (Whitridge,

BOX 6.2 SPECIFIC PLAY AFFORDANCES INDICATED IN RELATION TO OBSERVED, SITE-BASED, BIOPHYSICAL PROPERTIES IN WATSON'S (2017) STUDY OF CHILDREN AND URBAN WILD-SPACES

Site features play affordances and associated biophysical characteristics

• Entrance, start/finish	Picnic benches, toilets
• The roly poly hill	Rolling down, jumping down, rolling competitions
• Signage	Threshold between arriving and beginning exploration
• Trees	Climbing, jumping off, looking for insects
• Meadow	Open field, running, touching
• The river	Fishing, pebbles, wading, splashing, jumping off bank,
• The bridge	Dropping leaves, sticks, acorns into the water
• The horses	Feeding horses grass, stroking and petting at the fence
• Picnic/play area	Climbing, sliding, playing together on large basket swing
• Gates	Transitional points and threshold
• An old den	Natural debris, branches, bricks and stones, soft soil, insects
• The swamp	Mud
• The ditch	Water, leaves, jumping across
• The (shallow) river	Running up and down, jumping across, water, insects, tadpoles, mud, natural debris (branches stones) trees to build dens against
• The steps	Waiting point, threshold
• The climbing tree	Branches low enough for climbing
• The felled beech tree	Climbing, sawdust, fungi (King Alfred's Cakes)
• The mown field	Group games, flying kites, turning cartwheels, rolling down slopes

Notes: Some entries appear more than once, reflecting different biophysical properties at different locations. The affordances are derived from observing and recording children's experiences at the sites studied, and are by no means exhaustive. (Source modified from Watson, 2017).

2004; Massey, 1995; Watson, 2017). Children show sophisticated imagined trajectories as they explore nature sites, as affordances are realised deep person (child)-place interaction occurs. Such appear to offer particular stimulation to the child, drawing on their innate biophilic profiling.

Self/World Placings, Atmospheric Alignments and Biophilic Interactions

Study of children in urban wild-spaces reveals how children explore the properties of their biophilic worlds, engaging in natural rhythms and nature's cycles and responding to the biophysical properties of the wild-spaces. These are identified as 'atmospheric alignments' – cueing into environmental triggers, the rhythms and effects of natural phenomena and the biophysical properties which induce or support a favoured behaviour. For example, the feeling of 'the wind in my hair'; 'fun and muddy'; connecting with water, 'you can splash in it, you can drink it, it's healthy, and it's refreshing' (Watson, 2017, interviews). This study (see Box 6.3) used and refined filming as a technique for capturing the sophistication of interaction where the use of camera represents a way of seeing the world through the child's eyes and embodied reactions to give extremely rich sets of data. Filming records actions, reactions and narratives to allow an in-depth analysis of the links between child and environment which appear fundamental to understanding how the biophilic self is embodied in the individual through sensory cues and responses, experiencing and experimenting – 'Participants situate themselves as tacit explorers; thinking with

BOX 6.3 EXPLORING NATURE IN URBAN WILD-SPACES WITH YOUNG FILMMAKERS

- In a comprehensive study of school-children involving urban wild-spaces, Watson (2017) looked at 'person-place engagements'. Through observation, involving child-centric camera work and talking to the children, she was able to develop a typology which reflects the intensity of outdoor nature experience, evoking the term 'imagined trajectories'. Imagined trajectories relate to the connections the children drew between themselves, their actions and their imagination through 'atmospheric alignments' their bodies responding to and synchronising with the biophilic properties of places.
- The children sought out moments of high-quality engagement as 'pinnacle moments' of deep experience which typically involve physical and emotional intensity, e.g. rolling down the hill or running in shallow water, and 'pivotal moments of transformation' where the world is momentarily viewed from a different perspective.
- Watson was able to draw out these imagined trajectories as the children 'felt' nature at first hand when outside, due to the sensitive approach adopted and by analysing *'person-place-camera interaction'* with particular attention being paid to whole body movement, postures, the hands and the accompanying dialogue. She recognised 'pinnacle' and 'pivotal' moments of intensity and qualities that defined the experience. These

were highlighted from what Watson termed 'tellings performed in front of the camera'. For example, these might involve the camera capturing a non-verbal commitment between a small group of children (Prendville & Toye, 2007) and observed in their body language, hand gestures, the direction of their gaze, facial expression and spatial alignment or proximity to each other.

- Children recount their favourite moments as 'cut out and keep moments' which, although detached from the phenomena they are talking about, are reflexive expressions looking back on experiences and places where particular alignments have been intensively felt. These are related to favoured states and are closely aligned to personally effective ways of investigating and encountering the outside world. It is likely that they are pivotal in terms of the child's biophilic responses and ways that biophilic interactions are recognised by the body as the alignments are reasserted, sought out or reinforced through tactile engagement, the hands, turning to feel the wind in your face, touching or smelling.
- The camera acted as a conduit for these 'tellings' like an extension to the child's sensory and emotional experience, as it offered sophisticated actions such as panning, focusing in, turning the camera to feature on themselves, speaking in confidence, plus the children's roles as director, telling others what to do in front of the camera, or as actors giving the director feedback instructions. When accompanied by narrative 'speaking in front of the camera', it began to reveal what was happening.
- Filming captured the complexity and depth of the process in ways that narrative alone could not, and while it presents complexities for analysis, the data streams are rich in information and sensitive to the subtlety of the experience which could be expressed effectively to the outsider through interpretive sketches. In these, imagined trajectories drawings illustrated the blurring and integration of 'reality and irreality' in children's responses to environmental cues and triggers. Drawings allowed connections between participant and experience to be understood in ways that dialogue alone cannot, and collates incidences of pinnacle and pivotal moments, moments of intense physical interaction, as well as moments of misreading, demonstrating the mixing together of nature and the imagination.
- It was further possible to analyse the communicative behaviour involved in the film material regarding different aspects of discovery, exploration, reinforcement and retelling which described participants' voices and gestures performed in front of and when behind the camera. Through the use of these sophisticated techniques, a terminology was developed to describe how participants situated themselves in nature as 'Self/World Placings' involving 'Bio-Graphical Filming' (Watson, 2017)

FIGURE 6.1 Upside down. An interpretive drawing representing pinnacle and pivotal movements.

Source: Image credits: Sharon Watson.

touch, touching ground, touching and tracing key environmental cues and triggers in situ'. The hands and bodies and gestures externalise the intensity and nature of landscape interaction while the moments of silence suggest inner reverie or day-dreaming where 'nature is listened in to' as part of the 'bio-graphic' self. In Watson's study, the camera acts as a conduit and 'traces over environmental cues' that are

FIGURE 6.2 Sunset window: a cut out and keep moment for Zara watching changing colours in the sky; 'the best bit is when the sun goes down'.

Source: Image credits: Sharon Watson.

difficult to touch but which are integral to the experience. 'Camera movements follow tactile enquiry, as hands do'. It was thus possible to capture the intensity of the experience in nature to realise 'pinnacle moments', which offer a challenge to the senses and norms and appear to test the properties of the environment. For example, rolling down the hill and running in water evoke 'pivotal moments of transformation' where the world is momentarily viewed from a different perspective. The camera is a sophisticated lens where movements follow tactile enquiry, to reveal the depth of interaction and how it is cued from the biophysical properties of the environment being experienced through to its foundation as part of the individual's self. Seeing the environment through the lens of the camera, as indicated earlier, probably equates to the closest thing to viewing the world through the child's eyes and bodies and related senses, and demonstrates how they create their own space and dimension in nature (Watson, 2017; Lloyd et al., 2018).

Such auto-biographical filming thus gets to the heart of the biophilic experience and how it is rooted in the very physicality and physiology of the interaction demonstrated in the child's response and when outside, being influenced by the body situating itself within the environment, where it is indicated that innate biophilic interactions come to the fore. It seems that the direct outside experience on the child is critical, away from this direct contact (i.e. back in the classroom) things change, in that the experience may be mediated by other aspects; for example, ideas may morph into tales of fantasy, and other factors come into play as stories are constructed and 'the in-situ experience is distanced' (Watson, 2017). However, particularly profound nature experiences remain as strong influences; they endure as lasting and clearly meaningful memories or influences. Watson calls these 'cut out and keep moments'. These combine the intensity and importance of the pinnacle and pivotal moments to suggest a treasured experience which is cut out and kept for future reference and thus relived, reinterpreted and refined as part of self.

Creating Space for Nature and the Biophilic Self

Children need to create their own space in nature and they do this on their own terms (Thomas & Thompson, 2004; Bruce, 2008; Cobb, 1977; Watson, 2017). The process for experiencing it is laid down in childhood activities in contact with nature, and emphasises the importance of children being free to play in nature and related wild-spaces on their own terms. A range of studies confirm that attitudes to nature are developed in early childhood and endure into adulthood, where they remain as a positive experience in modern urban-based living (Derr & Rigolon, 2015; Yanez et al., 2017; Chawla, 2007). Childhood nature experience is believed to give the required confidence and process in the adult to gain a therapeutic response from natural interaction (Chawla, 2007), but it seems likely that childhood nature experiences need priming, cultivation and refinement just as other aspects of child to adult development do.

As such, a number of studies do stress the importance of contact with the natural world and are contained within a range of disciplines, such as environmental management, childhood studies and environmental education (for example, Derr & Rigolon, 2015; Rice & Torquati, 2013; Gill, 2014; Yanez et al., 2017). Some of these studies take a fairly focused definition of nature, but others adopt the idea of a more 'inclusive term pertaining to children's outside experience, including their interaction with the natural elements such as soil, wind, temperature, water, plants, and animals' (Yanez et al., 2017, p. 59). Yanez et al.'s work is interesting, as they adopt a novel methodology of using

FIGURE 6.3 Swinging: playing in nature forming an important moment of childhood experience.

squirrel puppets to enquire about children's values and definitions of nature. Puppets hold option cards which feature, in turn, a range of questions, i.e. placing the methodology within a play activity, but with space for follow-on responses, an approach which gave opportunities for the children to consider nature from their own specific experiences. The study was able to assess the biophilic credentials of children through a range of indicators and thus considered how children defined nature, but also showed that children understood the consequences of their actions on nature and how to behave in the natural environment (see Yanez et al., 2017).

In defining nature, children referred to 'plants including sticks and flowers, stars, familiar animals such as dogs, rabbits, kitties and chickens as well as wild animals including lions, cheetahs, zebras, animals not found naturally in the child's geographic area' (Yanez et al., 2017, p. 61). Pets were considered part of nature because they needed to be taken care of or fed. Wild animals were included as they 'need to live', even although some described them as 'mean and dangerous'. Similarly, those that included plants considered that they were important 'since you need to look after them', but also some responses emphasising the dangers of negative actions, such as dumping trash. From the results, it was clear that young children had a clear but personal construct of nature in relation to their experience, and that looking after nature was part of this understanding.

Research was carried out with pre-school children in urban and rural locations, but the researchers concluded that being located in an urban or rural district made little difference in responses. The authors also concluded that the definitions of

FIGURE 6.4 Den building: exploring the environment through play.

nature emphasise and reflect what they have learned through media and books, family, etc., in particular, being focused on things that needed care. As such, it excluded such elements such as water, weather or insects within their definitions, but may also be influenced by an understanding of death as a biological event around the age of six or seven (Slaughter, 2005), and thus what is animate and inanimate, what can die and needs caring for and what does not.

Research such as this, indicates aspects of how attitudes to nature are developed, how nature values are formed and articulated in children. As they grow up and are exposed to different ideas, concepts and opinions, they develop their own environmental identity which allows children to define themselves as part of nature including caring for nature as part of their role in the world (Chawla, 2007, p. 10). Children are seen to develop the emotional connection and values of nature before they start to think logistically about it, showing an understanding of existence, living, dying and survival, referencing the wider world and moral obligations, how the World and Universe are best cared for. These authors also emphasise the importance of exposure to nature-related experiences in young children (three to five years old), and that 'early childhood experiences may be foundational to developing a lifetime appreciation for nature as well as understanding of the relationship between their actions and nature's well-being' (Yanez et al., 2017, p. 64).

BOX 6.4 INDICATORS USED BY YANEZ ET AL. IN ASSESSING THE BIOPHILIC CREDENTIALS OF SCHOOLCHILDREN IN THE USA (NON-BIOPHILIC ASPECTS ARE IN ITALICS)

likes to play outside
likes to play inside
likes to dig for worms
does not like to dig for worms
likes to jump in puddles
does not like to jump in puddles
likes to watch birds
does not like to watch birds
likes to catch bugs and look at them
does not like to catch bugs and look at them
likes to watch animals like squirrels and rabbits
does not like to watch animals like squirrels and rabbits
likes to play in creeks and lakes
does not like to play in creeks and lakes
likes to play with sticks, leaves and pinecones
does not like to play with sticks, leaves and pinecones
likes to listen to birds singing
does not like to listen to birds singing
likes to look at the stars and moon at night
does not like to look at the stars and moon at night
likes to learn about wild animals
does not like to learn about wild animals

Notes: 1. Notice the range of biophilic indicators and their association with play or learning.
2. See Yanez et al. (2017) and also Rice and Torquati (2013).

Environmentally Focused Life Skills

The literature shows that outdoor learning offered on a regular basis has a range of benefits for children. For example, Tim Gill (2014) reviews the benefits of children's engagement with nature and observes that encouraging children to play freely in nature can improve their educational outcomes, and that the fun that comes from playing in nature is a crucial part of childhood. Prisk and Unsworth (2018), in their 'Muddy Hands' report, explored the same theme and analysed the impact of 'Outdoor Classroom Day', a global movement to inspire and celebrate outdoor learning and play (see Box 6.5); in the UK, Play England (see Lester and Maudsley 2007) adopted a similar theme exploring 'natural connections'.

These types of studies address the fact that children need routine exposure to the natural world and that such exposure provides important positive outcomes concerning educational achievement and social skills, but also address the need to develop environmental literacy as part of life skills (see, for example, Rigolon, 2012). In particular, they emphasise the role of the educational system and the importance of schools as influencers in promoting healthy environmental behaviours and responsibilities that are likely to persist through into later life (see Box 6.6). These skills involve the ways in which children gain experience and confidence in exploring and using nature and concern environmental literacy as a critical aspect of children's

BOX 6.5 OUTDOOR CLASSROOM DAY

- Outdoor Classroom Day is a global movement to inspire and celebrate outdoor learning and play. It was created to raise awareness about the importance of playing and learning outdoors, highlighting the benefits it brings to children, families and whole communities.
- The Outdoor Classroom Day movement emphasises that outdoor learning should be part of the school day for every child and argues for minimum targets, so that outdoor play becomes part of every child's everyday life.
- The campaign started in the UK in 2011 to spread the message that outdoors should be an essential part of every day.
- The campaign is led by the organisation Project Dirt, who have joined forces with leading education Non-Governmental Organisations and teams from Unilever's *Dirt is Good* brands to build local campaigns engaging with millions of children worldwide. Schools in over 100 countries have signed up.
- The movement supports the World Health Organisation's guidance regarding children and physical activity as well as the Sustainable Development Goals and the UN Convention on the Rights of the Child, which expect States to ensure children connect to nature.

Notes: 1. Based on Prisk and Unsworth (2018), see also www.outdoorclassroomday.com

BOX 6.6 THE BENEFITS OF OUTDOOR LEARNING

- Getting outdoors connects us to the places we live and the environments we will want to protect.
- Environmental stewardship and connection with place are strongly related to the amount of time we are immersed in nature as children.
- Children who are outdoors in nearby spaces often love their environment and grow up wanting to protect it.
- Getting outdoors results in better learning outcomes, across the board.
- Outdoor learning can create improvements across all academic disciplines.
- The benefits of outdoor learning and play last beyond early education.
- Outdoor play gets kids more active.

Notes: Summarised from Prisk and Unsworth (2018). See also Waite et al. (2016) and Wang (2017).

development within the formal educational process and advocate the adoption of environmental education programmes with schools, for example, Forest School or similar initiatives (see for example, Forest School Association (https://forestschoolassociation.org and the Forest Schools case study, in Chapter 9).

Conclusions

This chapter has explored various ideas relating to the urban child's environmental experience to establish the importance of contact with nature, explaining how they relate to biophilic interaction, including the notion of the biophilic self being evident in children from an early age. Through observations based on child-centric approaches, it is possible to see the importance of city wild-spaces to urban based children and how the qualities of the environment are investigated through bodily interaction and involve associated sensations of intensity, and that the ways that young children explore the natural world is largely based on embodied play related activities. The importance of these interactions is revealed through filming techniques which demonstrate the nature of such encounters and that these involve a bodily alignment with the natural world, in the form of atmospheric alignments and nature placings.

Several aspects are stressed:

- The importance of children being able to access the natural world.
- That nature experiences occur on children's own terms and that children be allowed to interact with nature and allied environments accordingly.
- In line with the Biophilia Hypothesis, it seems likely that biophilic engagement is inherent in children but will need priming, exploration and practise if

children are to confidently use their biophilic credentials in satisfying the need to access nature spaces.

- The development of environmental literacy and the ability to understand and experience nature first hand is a fundamental life skill that needs to be supported in children.
- The intensity of biophilic engagement is demonstrated in young children as a bio-graphical encounter involving embodied experiences and high-quality moments of sensation.
- It is thought that such positive child-related experiences and sensations are critical for later life regarding an individual's ability to access the positive restorative state that can be experienced through nature, biophilic placings and related circumstances.
- The importance of developing environmentally related skills in children is emphasised.

The development of environmental literacy, especially in terms of being mindful of nature's qualities, rhythms, fluxes and flows, may well depend on appropriate and adequate childhood exposure and the freedom for children to explore wild places on their own terms. Children are highly influenced by early exposure to nature and construct their ideas of what constitutes nature and how it needs to be cared for. Ideas concerning environmental literacy are discussed, involving the acquisition of environmentally focused life skills; studies emphasise that this should be part of the formal education process and demonstrate the importance of such skills in accessing the natural world in later life, including how the initiatives indicated also support other aspects of children's education and development.

Sharon Watson Biography

Sharon Watson trained as a landscape architect and led participatory design projects to facilitate user engagement and enhance access to nature in urban environments. She then worked as a forest schools leader in Birmingham, facilitating access to existing green spaces, leading outdoor play and nature based walks for children and families. Her PhD thesis developed an innovative participatory model, drawing on sensory ethnography and digital technologies, to engage child-led understandings of urban environments. She works at Birmingham City University as a research development and support officer.

References

Brady, E. (2002). Interpreting environments, *Essays in Philosophy* 3(1), 57–67. https://doi.org/10.2/JQUERY.MIN.JS

Bruce, T. (2008). *Cultivating Creativity in Babies, Toddlers and Young Children*. Oxford: Oxford University Press.

Bunkse, E. V. (2016). Feeling is believing, or landscape as a way of being in the world, *Geografiska Annaler: Series B, Human Geography* 89(3), 219–231. https://doi.org/10.1111/J.1468-0467.2007.00250.X

Chawla, L. (2007). Childhood experiences associated with care for the natural world: A theoretical framework for empirical results on JSTOR, *Children, Youth and Environments* 17(4), 144–170. www.jstor.org/stable/10.7721/chilyoutenvi.17.4.0144

Cobb, E. (1977). *The Ecology of Imagination in Childhood*. Columbia: Columbia University Press.

Coster, D., Gleeve, J. (2008). *Give Us a Go! Children and Young People's Views on Play and Risk Taking*. www.playday.org.uk

Derr, V. L., Rigolon, A. (2015). Participatory schoolyard design for health and well-being: Policies that support play in urban green spaces. In Freeman, C., Tranter, P., Skelton, T. (eds). *Risk, Protection, Provision and Policy*, vol. 12. Singapore: Geographies of Children and Young People, Springer, pp. 125–148. https://doi.org/10.1007/978-981-287-035-3_21

Duckett, R. (2009). Careless talk cost lives, *Refocus Journal* 14–16.

Ergler, C., Kearns, R. (2013). Children as explorers: Revealing children's views on well-being in intensifying urban environments. In Coles, R. W., Millman, Z. K. (eds). *Landscape, Well-Being and Environment*. Abingdon: Routledge, pp. 184–199.

Ewald, W. (2000). *Secret Games: Collaborative Works with Children 1969–1999*. Zurich: Scalo.

Gibson, J. (1986). *The Ecological Approach to Visual Perception*. Hillsdale: Lawrence Erlbaum Associates.

Gill, T. (2014). The benefits of children's engagement with nature: A systematic literature review, *Children, Youth and Environments* 24(2), 10. https://doi.org/10.7721/CHILYOUTENVI.24.2.0010

Gray, J. (2003). Open spaces and dwelling places: being at home on hill farms in the Scottish borders. In Low, S., Lawrence-Zuniga, D. (eds). *The Anthropology of Space and Place: Locating Culture*. Oxford: Blackwell Publishing.

Hinchcliffe, S. (1999). Cities and natures: Intimate strangers. In Allen, J., Massey, D., Pryke, M. (eds). *Unsettling Cities*. London: Routledge.

Ingold, T. (2008). Earth, sky, wind and weather. In Hsu, E., Low, C. (eds). *Wind, Life, Health: Anthropological and Historical Perspectives*. Oxford: Blackwell Publishing.

Johnson, C. N. (2000). The development of metaphysical thinking. In Rosengren, K., Johnson, C., Harris, P. (eds). *Imagining the Impossible*. Cambridge: Cambridge University Press.

Kallio, K. (2009). Between social and political: Children as political selves, *Childhoods Today* 3, 1–22.

Kaplan, R., Kaplan, S. (1989). *The Experience of Nature: A Psychological Perspective*. Cambridge: Cambridge University Press.

Kempen, H. J. G. (1996). Mind as body moving in space, *Theory & Psychology* 6(4), 715–731. https://doi.org/10.1177/0959354396064009

Kroh, D. P., Gimblett, R. H. (2007). Comparing live experience with pictures in articulating landscape preference, *Landscape Research* 17(2), 58–69. https://doi.org/10.1080/01426399208706362

Lester, S. (2011). Moments of nonsense and signs of hope: The everyday political nature of children's play, *International Play Association (IPA) World Conference, Playing into the Future-surviving and Thriving*. Cardiff University Press.

Lester, S., Maudsley, M. (2007). *Play, Naturally: A Review of Children's Natural Play*. London: Play England.

Lloyd, A., Gray, T., Truong, S. (2018). Seeing what children see: Enhancing understanding of outdoor learning experiences through body-worn cameras, *Journal of Outdoor Recreation, Education, and Leadership* 10(1), 52–66. https://doi.org/10.18666/JOREL-2018-V10-I1-8192

Louv, R. (2005). *Last Child in The Woods: Saving Our Children from Nature-deficit Disorder*. Chapel Hill: Algonquin Books.

Lund, K. (2012). Landscapes and narratives: Compositions and the walking body, *Landscape Research* 37(2), 225–237. https://doi.org/10.1080/01426397.2011.651111

Lury, K. (2010). Children in an open world: Mobility as ontology in New Iranian and Turkish cinema, *Feminist Theory* 11(3), 283–294. https://doi.org/10.1177/1464700110376279

Macpherson, H. (2010). Non-representational approaches to body – Landscape relations, *Geography Compass* 4(1), 1–13. https://doi.org/10.1111/J.1749-8198.2009.00276.X

Massey, D. (1995). The conceptualization of space. In Massey, D., Jess, P. (eds). *A Place in the World?* Oxford: Oxford University Press.

Moore, R. C. (1987). Streets as playgrounds. In Moudon, A. V., Appleyard, D. (eds). *Public Streets for Public Use*. New York and Wokingham: Van Nostrand Reinhold.

Mustapa, N. D., Maliki, N. Z., Hamzah, A. (2015). Repositioning children's developmental needs in space planning: A review of connection to nature, *Procedia – Social and Behavioral Sciences* 170, 330–339. https://doi.org/10.1016/J.SBSPRO.2015.01.043

Nell, M. L., Drew, F. (2022). Five essentials to meaningful play, *Naeyc Originally Posted on Families.Naeyc.Org*. www.naeyc.org/our-work/families/five-essentials-meaningful-play

Nemeroff, C., Rozin, P. (2000). The nature and function of sympathetic magical thinking. In Rosengren, K., Johnson, C., Harris, P. (eds). *Imagining the Impossible*. Cambridge: Cambridge University Press.

Noe, A. (2004). *Action in Perception*. Cambridge: MIT Press.

Ohta, H. (2001). A phenomenological approach to natural landscape cognition, *Journal of Environmental Psychology* 21(4), 387–403. https://doi.org/10.1006/JEVP.2001.0233

Piaget, J. (1973). *The Child's Conception of the World*. St. Albans: Paladin.

Prendville, F., Toye, N. (2007). *Speaking and Listening through Drama 7–11*. London: Paul Chapman Publishing.

Prisk, Unsworth (2018). From muddy hands and dirty faces…to higher grades and happy places, *Muddy Hands Global Report*. www.playscotland.org/resources/muddy-hands-global-report-november-2018/

Reding, C. (2009). The sounds of leaping: Exploring relationships between music and movement, *Refocus Journal* 9(Autumn).

Rice, C. S., Torquati, J. C. (2013). Assessing connections between young children's affinity for nature and their experiences in natural outdoor settings in preschools, *Children, Youth and Environments* 23(2), 78–102.

Rigolon, A. (2012). A greener future: The active role of place in enhancing ecological literacy in children, *Journal of Architectural and Planning Research* 29(30), 181–203.

Rissotto, A., Giuliani, A. M. (2006). Learning neighborhood environments: The loss of experience in a modern world. In Spencer, C., Blades, M. (eds). *Children and Their Environments: Learning, Using and Designing Spaces*. Cambridge: Cambridge University Press.

Singer, D., Singer, J. (1992) *The House of Make-believe: Children's Play and the Developing Imagination*. Cambridge: Harvard University Press.

Slaughter, V. (2005). Young children's understanding of death, *Australian Psychologist* 40(3), 179–186. https://doi.org/10.1080/00050060500243426

Spencer, C., Blades, M. (2006). *Children and Their Environments: Learning, Using and Designing Spaces*. Cambridge: Cambridge University Press.

Thomas, G., Thompson, J. (2004). A child's place: Why environment matters to children, *Green Alliance/Demos Report*. www.demos.co.uk/files/AChildsPlace.pdf

Thrift, N. (2000). Afterwords, *Environment and Planning D: Society and Space* 18, 213–255. https://doi.org/10.1068/d214t

Thwaites, K., Simkins, I. (2007). *Experiential Landscape: An Approach to People, Place and Space*. Abingdon: Routledge.

Waite, S., Passy, R., Gilchrist, M., Hunt, A., Blackwell, I. (2016). *Natural Connections Demonstration Project, 2012–2016: Final Report* (Natural England Commissioned Reports). London: Natural England.

Wang, J. (2017). The wonders of outdoor education, *SingTeach*. https://singteach.nie.edu.sg/wp-content/uploads/2018/01/SingTeach-Issue-63-1.pdf

Ward, C. (1978). *The Child in the City*. Harmondsworth: Penguin.

Watson, S. (2017). *Exploring Nature in Urban Wildspaces with Young Filmmakers*, (Unpublished PhD Thesis). Birmingham: Birmingham City University.

White, R., Stoecklin, V. L. (2008). Nurturing children's biophilia: Developmentally appropriate environmental education for young children. Collage: *Resources for Early Childhood Educators*. www.whitehutchinson.com/children/articles/downloads/nurturing.pdf

Whitridge, P. (2004). Landscapes, house, bodies, things: Place and the archaeology of innuit imaginaries, *Journal of Archaeological Method and Theory* 11(2), 213–246.

Woolley, J. D. (1997). Thinking about fantasy: Are children fundamentally different thinkers and believers from adults? *Child Development* 68(6), 991–1011. https://doi.org/10.1111/J.1467-8624.1997.TB01975.X

Yanez, R. E., Fees, B. S., Torquati, J. (2017). Preschool children's biophilia and attitudes toward nature: The effect of personal experiences, *The International Journal of Early Childhood Environmental Education* 5(1), 56–67. https://naaee.org/eepro/research/database/preschool-childrens-biophilia-and

7

URBAN FOOD GROWING AS A BIOPHILIC PATHWAY

Introduction

This chapter explores the specific features of urban food growing and community gardening as a biophilic pathway, examining how human-nature interactions are manifest in urban food growing initiatives and the ways that the movement is presenting itself. As an activity, urban food growing concerns a level of engagement and interaction which demands particular attention to the environment involving an 'attachment to the land', activities being dictated by cultivation, the soil, the seasons, observing and responding to the weather. Accordingly, there is a synchronicity between parameters relating to biophilia and the parameters relating to cultivation, tending plants and the harvesting of food, with the additional satisfaction of preparing and consuming produce. As such, it embraces a sophisticated suite of outputs which involves highly positive ecological and social interaction delivered through the activities associated with growing and sharing the same activities with other like-minded individuals.

It might initially be thought that urban food growing represents a divergence away from the main body of discussion around biophilia, but in fact, as an urban society, we are as much isolated from food production as we are from nature, and urban food growing can be explored as an alternative, or perhaps more accurately a complementary, pathway to achieving biophilic satisfaction which, in the ways that it is now being considered, appears to have unique properties. These include the associated physical activity, haptic connections and a mindful approach to environmentally focused tasks of cultivation, which concern the generation of strong positive flows through the body, underpinned by profound multi-sensory experiences, in combination with the involved social and cultural aspects.

DOI: 10.4324/9781003099758-8

Food Growing and the Urban Agenda

Food as a socio-ecological system has, of course, been intrinsically linked with the lives of cities throughout history, with food production being an integral part of the dynamic, within and outside of the city walls or boundaries. Most of all, food mediated people's interactions with landscapes and cities, and it is this which still underpins the current urban food growing movement. However, this dynamic, despite being fundamental to human existence, has not maintained its presence in any major way as part of mainstream urbanisation. As the last urbanite generations have gradually become disconnected from the land, urban food growing seems to offer a way of reminiscing with our evolutionary paths, reconnecting people with the land, their food, their culture and tradition. As Carolyn Steel states in her book Sitopia (Steel, 2020),

> Food shapes, connects and flows through our lives and the world in so many different ways, linking our past, present and future.
>
> *(Steel, 2020)*

Food growing is largely divorced from urban planning, and has not been afforded a place under the pressures and models of urban expansion. Under these circumstances, there is no longer any meaningful connection between the growing of food and its provision; this disconnection from the land has broken one of the most basic of biophilic/survival linkages. This disconnection is particularly evidenced in children, with serious consequences for their diets (Wilson, 2022).

The globalisation of food production in its current form, i.e. via an industrialised agriculture industry, is of course a response to the demands of feeding the world and securing global food security; urban food growing, on the other hand, and especially in the context investigated here, is as much about restoring the link between people and the land as it is about growing produce and, like other aspects of the urban environment, is being explored by a range of disciplines and on a cross-disciplinary basis, including planning, landscape, architecture and urban design. The movement also reflects a certain degree of dissatisfaction with conventional urban landscape systems, in that they fail in offering deeper engagement or linking people to their environment in ways that urban food growing can. Accordingly, the system concerns both the practical aspects of food production but also cultural aspects, being underpinned by this basic tenet, i.e. associated with maintaining opportunities to work the land, produce food, self-sufficiency, restoring lost connections and realising new ones through the process of cultivation and community endeavour. The dynamics of the system, the benefits to society, community and to the individual are thus being re-assessed in the context of modern urban living, including the social, health and environmental agenda, while also helping to address concerns about diet and food literacy.

Accordingly, there has been a wealthy body of research demonstrating that urban food growing provides multifunctional opportunities for individuals, communities and cities, often discussed as benefits in the form of ecosystem services (Langemeyer et al., 2018; Speak et al., 2015), as integral part of green infrastructure (Breuste, 2010; Costa et al., 2016) and in contributing to urban resilience and sustainability (Viljoen, 2005). However, individual motivations, while underpinned by the production of food, typically for self-consumption, offer a great number of opportunities and benefits. As suggested, it is not the production of food that is always the primary motivation, although importantly, this does provide the context. Involvement may be related to improving health, addressing social isolation or be located in an educational setting, where the experience of growing food acts as an activity and pathway in which personal goals are realised.

Current explorations are wide-ranging in their remit and reflect this diversity of outputs, including recreation, relaxation and escapism, opportunities for connecting with nature (Dubbeling, 2011), spaces for connecting people, social interaction and social innovation (Delshammar et al., 2016), helping to tackle poor social contact and loneliness. Other investigations are set within an educational context (Walter, 2013) and in the provision of therapeutic environments, exploring how food-growing activities influence human health and well-being, including improving self-esteem and mood, reducing depression, anger and confusion, physical health and fostering healthy diets (Alaimo et al., 2008; Litt et al., 2011; Wood et al., 2016).

Exploring Urban Growing Models

This renewed interest in urban agricultural systems is driving the emergence of multiple urban spatial growing models. These are a response to various issues, including the inadequacy of and related frustration with traditional schemes in meeting demand (and lengthy waiting lists), development pressures and the slow adaptation of planning systems to the most recent environmental, social, economic and political challenges (Caputo et al., 2016). However, they are also set within the context of green activism, and related socio-political engagements which concern various rights to access and cultivate the land (Hardman et al., 2018; Ioannou et al., 2016) and involve different ways of expression in different configurations and urban contexts (Costa et al., 2016; Coles & Costa, 2018). Accordingly, they show substantial diversity in the ways that they are being explored and the different socio-ecological situations they occupy or complement, including the new opportunities they embrace. They also reflect a yearning for and a way to explore meaningful human-nature interactions and greater nature relatedness (Schunko & Brandner, 2022, Steel, 2020).

> Gardening is the opposite of consumerism: one must be active, engaged, patient, observant, empathetic and above all in sync with nature.
>
> *(Steel, 2020)*

FIGURE 7.1 Brooklyn Grange is a 5.5-acre organic urban rooftop farm in New York City, growing vegetables and honey for local restaurants, markets and community-supported agriculture. The farm is located on three rooftops in Brooklyn and Queens.

Source: Image credits: Rhododendrites CC BY-SA 4.0 commons.m.wikimedia.org

The result is that the different niches and models being explored are wide ranging at both city and local scales, in new and conventional situations, including the conversion of rooftops into social enterprise farms (Figure 7.1, www.brooklyngrangefarm.com) and many other concrete plots into productive spaces. These involve developing city-wide ecologies that connect neighbourhoods through the power of food growing and alternative urban agricultural systems, creating alternative landscape systems which are underpinned by the patterns, structure and aesthetics of food (Viljoen, 2005; Viljoen & Bohn, 2009).

Caputo et al. (2016), in considering the range of initiatives, suggests that these urban growing explorations can be grouped broadly according to three socio-political dynamics:

1 **The legal or illegal reclaiming of unused urban space**, previously under another use, in the form of growing; these spaces are usually disconnectedly located across the city and can be presented at a range of scales, sometimes occupying marginal or very small spaces, in soil conditions that may or may not be permeable or healthy land.

2 **The integration of growing spaces in buildings**, not only as a form of improving urban environmental conditions, and potentially the building itself, but also as strategy to make available such spaces for growing by its residents or workers; this may involve the inclusion of spaces for growing in balconies, terraces, walls, and rooftops.

3 **The integration of such spaces in the urban fabric through the planning system**, a situation embraced by local governments as part of green infrastructure strategy or other planning integrated (e.g. in conjunction with accessibility and cycle infrastructure) or regulated approaches, and as a recognition of their value as social and spatial tools to engage the city in community practices.

Accordingly, as manifest, urban food growing systems demonstrate a variety of contrasting socio-spatial associations and environmental niches, ranging from the formality of rigid structures to freestyle spatial experimentations that transcend conventional urban systems and are considered more integrative of a number of audiences and users (Costa et al., 2016; Scott et al., 2018). These involve new expressions of action concerning food growing in multiple circumstances, but also raise issues of how we interface with the environment, how and where such connections can and perhaps should be made in creating urban environments which are accessible in a variety of ways, embracing flexibility in the type of interactions, typologies and niches (new and old) available (Figure 7.2). Thus,

FIGURE 7.2 Edible Eastside in Birmingham (before its dismantling in 2016). Pop-up, temporary community garden located in a derelict brownfield. The initiative explores a new typology that is underpinned by an aesthetic that considers urban regeneration and green activism through food growing and the appreciation of post-industrial features.

they concern wider issues associated with planning policies and landscape strategies as whether they are fit for purpose (Coles & Costa, 2018). This is especially so, since the emerging approaches and the diversity of typologies seem to offer advantages from spatial and user perspectives which can be applied at any scale to fit a range of spaces and circumstances on both a temporary and permanent basis, including parking lots, rooftops, balconies, vacant land, brownfields, streets and urban parks.

With a number of key authors and initiatives advocating for an urban food model to be developed as a more mainstream aspect of city development (see, for example, The Carrot City, [Box 7.1] and EU COST Actions on Urban Agriculture and Allotment Gardens in Europe, 2012–2016), examples around the world provide strong support and continue to raise discussions around alternative food systems and the pressing need to include urban food growing in the debate on sustainable development, including the reintroduction of food growing as a more mainstream activity in cities.

These include strong visions regarding preferred lifestyle choices; for example, Rob Hopkins (www.robhopkins.net/the-book/), in writing about how our current modern lifestyles are impacting on our individual and collective imagination, envisions how city places might be intrinsically linked with food and daily routines, thus to underpin our human condition, our existence and the everyday, posing the question 'what if?':

BOX 7.1 CARROT CITY

- Carrot City is a research initiative that explores how design at all scales can enable the production of food in the city. The initiative collates a wide range of examples and aims to disseminate ideas and knowledge about best practice featuring a large number of case studies which show the diversity and range of initiatives.
- As such, it offers an extensive database which can inform practice, including conceptual and realised examples.
- Focus is on the need and desire to grow food within the city, to supply food from local sources, including the contribution of architecture, landscape design and urban design, thus helping to enable sustainable food production and helping to introduce urban agriculture to our cities.
- It investigates the impact that agricultural issues have on the creation of urban spaces and buildings, how an increasing interest in food growing in the city is changing urban design and built forms.
- Examples are wide-ranging and address the entire spectrum of urban situations and locations, such as rooftop school gardens, community

orchards, rooftop greenhouses, vertical farms, positive eating, agro-parks and agro-housing, the edible terrace and urban agriculture as green infrastructure.

- Examples are presented under five categories which reflect different scales and opportunities.
  - Imagining the productive city
  - building community and knowledge
  - redesigning the home
  - producing on the rooftops
  - components for growing

(See www.carrotcity.org)

> I wake my kids, get them dressed and fed and accompany them to school – a walk that takes us through shared gardens with a diversity of food crops, including young ruby chard whose deep red leaves radiate like stained glass caught in the brilliant sun of this late spring morning. The streets are quiet, due to sparse motorized traffic, and they are lined with fruit and nut trees in early blossom. The air smells of spring. Each bus stop we pass is surrounded by a garden on three sides, part of the Edible Bus Stop network that now includes most bus stops across the United Kingdom. Anyone can graze while they wait for the bus.
>
> *(Hopkins, 2019)*

BOX 7.2 FOOD GROWING AND EMBODIED MULTISENSORY INTERACTIONS – THE CASE STUDY OF WALSALL ROAD ALLOTMENT GARDENS

Between 2012 and 2018, we visited a number of urban growing places in Europe. Walsall Road allotment gardens in Birmingham was one of them where we had the opportunity to interview and talk to a number of gardeners on site. The Walsall Road allotments are a typical British allotment garden divided into individual plots. Each plot is personalized regarding its layout, the produce growing, the decorative elements such as flowers, porches, fences and pathways (Figure 7.3) and the more functional structures such as sheds, rainwater harvesting and storage containers, greenhouses and composting bins, most of the time made of recycled and reused materials.

One of its interesting characteristics is that of multifunctionality underpinned by features that promote social interconnectedness and foster diversity of uses, which support a 'garden community'. Examples of this include being run by a garden committee and offering several communal areas, a community centre/pavilion, grassed edges with mature trees, and opportunities for seating, fairs/markets, gatherings and barbecues. Some of its users

FIGURE 7.3 Walsall Road Plot 30 'Mentalot' – the plot is a place for exploring self-identity through personalisation of spatial and decorative elements.

described it as a 'fantastic site, like a park' and 'probably the best allotment site in Birmingham' (Figure 7.4). At Walsall Road, many highlighted social bonds within smaller groups, e.g. chatting, healthy banter, sharing plant and growing knowledge, exchanging seeds and produce, watering each other's produce when needed and having tea together, describing the relationship as friendly and as an extended family.

Conversations with gardeners (see below) were often dominated by referencing evocative and profound bodily physio-emotional exchanges and to the associated multi-dimensional sensory attributes, suggesting a flow between the body and the growing activities. This 'flow experience' seeks to understand how people feel when they are in a high state of enjoying themselves, and the concept describes a state of mind related to that of joy, creativity and full involvement, in which the person is totally absorbed, to the extent that they feel in transcendence, to the point of forgetting time, fatigue and everything else but the activity and where problems seem to disappear (Csikszentmihalyi et al., 2005; Nakamura & Csikszentmihalyi, 2009).

Interviews substantiated the multiple nature of the gardening processes, capturing the experience in words but concerning how the body is aligned to undertake different tasks, the diversity of these related tasks and haptic associations involved in preparing the soil, sowing, watering, weeding, caring and watching plants grow, harvesting and finally the pleasure of eating your own produce.

> The smell of water, the hum of a bee, the directional bend of a plant stalk mattered.
>
> (Wilson, 1984, *p. 189*)

FIGURE 7.4 Walsall Road, the interface between communal spaces and the plots.

The planning that goes into the growing activity, i.e. what to grow, when and where, and the bodily experience associated with the cycles of growing and harvesting demonstrates that there is also a synchronicity with the seasons and a natural order of things, including, of course, an alignment with seasonal food eating and the exceptional flavour, colour and texture that comes with it, appreciation that is developed and grows over the time spent in the allotment.

Gardeners at Walsall Road clearly showed a strong emotional response induced by the various processes of cultivation, and that are linked with the certainty of the satisfaction of their own making and of drawing positive outcomes.

> *This is a great place to remove stress.* ***Digging is*** *wonderful if you are all uptight with problems at work and you dig for half an hour and it's like the stress is going into the soil away from you. And this is a great stress buster.* Teresinha
>
> *I find this very therapeutic because I can lose myself there. Even if it's only just* ***weedy,*** *if I can weed for 2 or 3 hours put me minds gone, if you know what I mean. No worries at all, so I find it very therapeutic.* Trevor
>
> *You need to be fit. I see a lot of people come because maybe they are overweight or the doctor says that they should get exercise and try gardening, but really it's quite heavy work so, you need to be fit to keep on top of it. You have to do a lot of weeding, a lot of digging so you need*

to be fit really but no more so if you took up a sport or anything else really. Pat

Today, I'm doing a fire, burning rubbish. But each time is different really. The trees need pruning, so I come down and prune the trees and they need spraying as well. And then the seeds need setting as well and the plants need putting in and all that. Trevor

When I am down here, I am in a completely different world. People can walk up and I wouldn't know they were there . . . I am miles away with what I'm doing, I am concentrated too much. Bill

If you can get your stove shed, you need the bees to pollinate your food, because without the bees you're not gonna get any food. If you can grow your flowers that bring the bees in and the butterflies and the yellow other flies in to pollinate your runner beans or your peas and else, and you know, but if you just gonna grow and grow without helping the bees, you know, you are not gonna get anything. Bill

Teresinha, one of the gardeners who first started growing a plot at Walsall Road because she was a newcomer and wanted to meet other people, grows linseed and dye plants from which she makes her own jumpers and dyes for the wool and fabrics that make her own clothes.

Cloe and Bill deliberately plant floral borders in their plots not only because their colours are aesthetically pleasing to the eye but also because they wanted to attract more bees and butterflies to the plot.

Steve puts bird feeders out and gets all different birds, and created a small pond with an old bath in his plot to attract frogs and newts.

The Well-Being Properties of Growing

The bodily encounters and actions involved in cultivation, of touching the soil, the seeds, weeding and harvesting, describe 'haptic interactions of food growing practices'; the well-being properties of growing are thus 'embedded in its materiality' (McGuire et al., 2022) and the process employs strong sensory and perceptive dimensions (Figure 7.5) that also appear to connect our modern lives culturally to our ancestors and, for many of us, a connection to memories of childhood. These responses are well demonstrated via interviewing those engaged in food-growing activities, and identify the nature of the positive experiential flow influenced by consecutive moments of physical immersion and mental detachment. This includes enjoying the physical exercise associated with the task of cultivation

FIGURE 7.5 Growing practices in action at Walsall Road – digging, dealing with manure, burning dry vegetable matter.

FIGURE 7.6 Chelas Valley Allotment Park (Parque Hortícola do Vale de Chelas), Lisbon. Food growing integrated within the urban fabric in close proximity with socially vulnerable communities.

and the reward or achievement of producing food (i.e. the tasks are goal-driven), anticipating the moment to take the produce home.

Clearly demonstrated is that when it comes to urban food growing, it is sensory stimulation and exploration which acquire particular significance in contributing to a positive experience and enacting embodied restoration, particularly since our modern lifestyles have led us to sensory deprivation (Tilley, 2006).

It seems that the materiality and haptics of growing form a means of engagement which can operate at virtually any scale and that the immersive nature of the experience and the consequent mental detachment can be manifest in different situations, including the setting up of temporary (pop up) growing installations; this makes the urban food system a particularly flexible one in terms of supporting mental and physical well-being (Figure 7.6). As an urban activity, it is the potential of the system to be applied in such a variety of ways and in different contexts associated, that make urban food growing activities so flexible and highly relevant to the discussion on environmental interaction.

Urban Food Growing as a Community of Practice

Many urban locations provide strong opportunities for establishing connections between individuals and nature, and the practices of growing and the activities involved might be regarded as a 'community of practice' with associated social implications (and other aspects) which occur as a default, i.e. in providing opportunities for developing social connections through place making and through the practice of growing (Noori & Benson, 2016; Kurtz, 2001).

Social connectedness, addressing social isolation and loneliness are seen as key reasons why people decide to join a community garden and is demonstrated throughout cases (Kingsley et al., 2009). For urbanites living in a flat and not having a private garden, or being unable to use common facilities to grow food, access to an urban growing space can provide a particularly significant opportunity to connect with others (Figure 7.7). For example, activities focused around a food growing location involve opportunities to share knowledge about the weather, the soil and nutrient cycles, the seeds, the plants and resources. It is this mix of activities, connections and place identity which underpin the specific features of food growing, i.e. as a biophilic pathway, forging social relationships and social inclusion, helping establish environmental literacy, building stronger social networks and nature-based experiences, working alongside others who share the same views.

> We can forge new social connections and friendships and evolve a closer relationship with nature that doesn't increase our footprint on the globe – all of which bring us back to food.
>
> *(Steel, 2020)*

FIGURE 7.7 Bill and Steve (2016) in Steve's shed. Allotment gardening can be considered as a community of practice in which family, friendship and other social activities can form a major part.

Socio-Ecological Connections and Nature-Associated Attributes

Within the urban growing community, place making and political fulfilment are seen as being strong drivers for connecting communities focused around common goals and strong complementary motivations which foster socio-ecological experimentations, local ecological knowledge and knowledge exchange (Camps-Calvet et al., 2016). The specifics of individual growing experiences reflect personal identity and values, express a sense of ownership, the need to be sustainable and encouraging others to do the same. Such collective action plays an important role in building city, community and nature resilience, supporting biodiversity and ecological functions and attitudes towards environmental activism (Camps-Calvet et al., 2016; Andersson & Barthel, 2016; Andersson et al., 2014).

Socio-ecological diversity is also reflected in the construction and organisation of a plot or a growing space or wider food growing landscape. For example, many allotment gardens in the UK can show a distinctive and somewhat random character, being the subject of the requirements and high diversity of crops grown, individual responses and the practical aspects of cultivation. However, if allowed to realise their full potential as community gardens, they more fully reflect community identity and an aesthetic of sustainability and recycling (e.g. through the use of recycled materials employed in greenhouses and sheds) and a landscape that allows for states of overgrowth and wilderness as part of nature, expressions of lifestyles, values, political and ecological ideologies. When developed to their full potential, they can form distinctive meaningful landscape places which are 'owned' and defined by their respective 'garden community', with an underpinning management plan and structure (Costa et al., 2016). Unfortunately, the lack of mainstream attention on urban food growing means that the aesthetics of food growing places is barely explored, understood or supported in realising the sophistication that could be achieved.

Despite different arguments on their aesthetic reality (Costa et al., 2016), urban growing spaces embody nature-associated attributes that are mediated through human experience and engagement and which represent and symbolise different positions in respect of biophilic connections; how, when and under what circumstances connections are made, the medium of the experience and also how they are underpinned in social, educational, aesthetic, health or other contexts. In the case of urban food growing, this concerns a strong link with the land and the adoption of a stewardship approach, and in the current trajectory of urban food growing, involves multi-connections that are profiled through different examples, but of course conform to individual need. Multi-deliverables are the norm, and working together through a community of practice, the socio-ecological functioning is restored or enhanced in multiple ways across scales and locations, including urban biodiversity performance. Pertinent to discussion is that relationships with nature and biophilia can be and are expressed in different ways and concern different pathways which involve different mixes of circumstances, but these different expressions still motivate pro-active approaches regarding the environment, nature and the adoption of more sustainable behaviours (Figure 7.8), and can be particularly powerful in fostering stewardship (Hensley, 2015).

FIGURE 7.8 The environmental aesthetics of the allotment. Bug hotel, rainwater harvesting and recycled materiality.

Continuous Productive Urban Landscapes and Integrated Food Growing Systems

The idea of 'Continuous Productive Urban Landscapes' (CPUL), developed by Viljoen & Bohn (see Viljoen, 2005; Viljoen & Bohn, 2009), considers the concept of an urban-wide food growing system as an interlinked network of places both within the city and between the city and the countryside. As a concept, it addresses issues concerning a sustainable approach to food growing, involving food growing as an integrated activity alongside other urban functions, but particularly regarding open space planning. As such, CPUL can be discussed regarding ideas of urban green infrastructure, its socio-ecological features and its landscape qualities, plus the characteristics that urban food growing systems can display regarding biophilic, haptic and health connections.

Viljoen and Bohn (2009) discuss the system 'as a strategy for the coherent integration of urban agriculture into urban space planning' (p. 50) and where 'Green corridors provide a continuous network of open space contained footpaths and cycle ways. Fields for urban agriculture, playing spaces and other leisure activities are located within the network and serve adjacent built up areas' (p. 51).

The concept is underpinned by a range of precedents and explorations regarding economic and aesthetic criteria, particularly an analysis of urban agriculture in Cuba, but is clearly more generally endorsed by the role that urban food growing demonstrates, i.e. as a multifunctional component of the city and its power in connecting people to their environment, instilling environmental awareness and promoting stewardship.

Viljoen and Bohn (2009) explore the concept across a number of themes, including the aesthetics of such an approach (Box 7.3) and its credentials regarding sustainability, including:

- The idea that urban agriculture can be read as ornament.
- The debate on infrastructural urbanism.
- Urban landscape as a major contextual and lifestyle component for the design of a sustainable, contemporary city.
- Addressing the ecological footprint of buildings.
- The economic advantages of locating fruit and vegetable production adjacent to urban areas.
- A new and inclusive approach to urban space planning to justify the use of inner-urban open space for non-building activity.
- The example and analysis of Cuba's urban agriculture programme, including economic and aesthetic considerations.

(summarised from Viljoen & Bohn, 2009, see original for full details)

BOX 7.3 THE AESTHETICS OF CPUL (BASED ON VILJOEN & BOHN, 2009)

The aesthetics of the CPUL system can be defined by a number of factors, including:

- The visual patterns resulting from the functional layout.
- Enclosure – strong visual presence, which impacts on how the space is read.
- Multiple use – specific uses beyond the growing of food, such as outdoor classrooms, picnic sites, sales points and health advice centres.
- Shared visual facility – shared visual resource, linking the public to natural systems that sustain life, connections between residents and farmers, provide a sense of open space.
- Linking device – growing beds provide physical and optical links across fields and larger territories, assist in unifying and giving coherence to potentially disparate parts of the urban fabric.
- Sculptural quality – the horizontality of growing beds accentuates local topography.
- Incremental occupation – related to the temporary or incidental land uses and patches of open ground used for such events as wedding receptions and circuses.

Notes

1) Compiled from ideas discussed by Viljoen and Bohn (2009).
2) For full information and original wording, see original source.
3) In the Cuban example, urban agriculture accounted for 60% of vegetable production, and it was estimated that the approach could make a significant contribution to fruit and vegetable production in other regions.

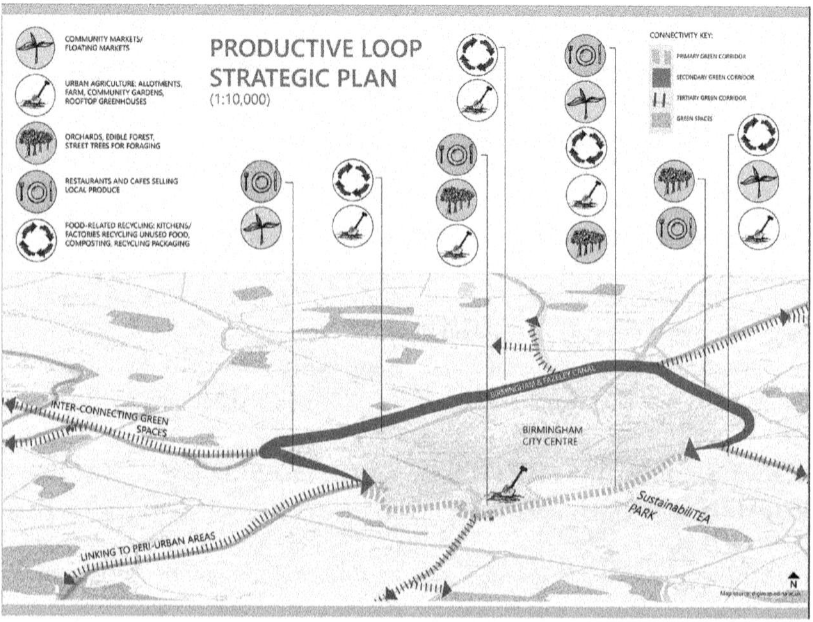

FIGURE 7.9 The Productive Loop for Birmingham City Centre. The concept identifies a number of place characteristics and functions that embrace different levels of physical and mental activity, relaxation or exertion and give a substantially more distinctive, purposeful and connected landscape character to the city centre by integrating a number of landscape/food-related niches.

Source: (see main text for more information). Image credits: Kerryn Hooper.

Viljoen and Bohn (2009) subtitle their discussion 'essential infrastructure and edible ornament', and the CPUL concept has been investigated in a number of situations, including how the philosophy involved might underpin design thinking within a landscape planning context. Such exploration was carried out by postgraduate landscape architecture students at the Birmingham School of Architecture and Design (Birmingham City University, UK), in relation to the landscape and functionality of Birmingham City Centre (Figure 7.9). These demonstrate that the ideas of connectivity, through both spatial patterns and the food growing theme, do result in a critical appraisal of opportunities to deliver an effective landscape structure capable of providing the multi-benefits associated with food growing systems and introduce additional meaning and functionality into the landscape beyond its commercial activity, i.e. acting as a mediating component.

Kerryn Hooper, for example, visualises the idea of a 'Productive Loop' for the City Centre (Figure 7.9) which incorporates a diversity of places linked by the context of food growing and consumption. Kerryn identifies a number of place characteristics and functions, ranging from growing, to buying, to socialising and eating places, that embrace different levels of physical and mental activity, relaxation or exertion and give a substantially more distinctive, purposeful and connected landscape character to the city centre by integrating a number of landscape/food-related niches, including:

- Urban agriculture allotments,
- Floating markets,
- Farm community gardens,
- Rooftop green houses,
- Orchards and edible forests,
- Street fruit trees for foraging,
- Restaurants and cafes selling local produce,
- Food-related recycling kitchens/factories recycling unused food,
- And composting.

Conclusions

This chapter has discussed ideas concerning urban food growing to explore how it adds to our understanding of environmental engagement and the application of urban food growing as a complementary biophilic pathway, focusing on the ways that it facilitates socio-ecological connections. It is clear that urban food growing initiatives have features that provide strong connectivity to the land, environment, communities and personal well-being, and are a highly legitimate urban land use.

Being explored in a multidisciplinary context, urban growing initiatives are particularly diverse in their application, addressing the complete range of urban niches, including built forms. In particular, it is the features of food growing, how they can introduce additional 'socio-ecological power' into urban landscape systems, together with their versatility of application, that needs to be considered in city environmental strategies. These include the ways that food growing stimulates strong sensory responses involving mental restoration and social interaction, with strong motivation being provided by the activities as well as the physical effort, and positive impacts on mental health.

The debate on urban food growing focuses on the different dimensions that the systems offer, including how they provide connection to the environment, nature and natural process and involve a stewardship approach, while the rapidity with which schemes can be developed, including temporary pop-up gardens in virtually any circumstances, demonstrates the potential of the system. Urban food-growing models, including urban agriculture, have also been analysed regarding their aesthetic characteristics and how urban agricultural field patterns can help produce a balanced urban landscape acting in harmony with other considerations as part of urban infrastructure, and how such systems help moderate the harsher aspects of urban living and the dominance of built forms.

References

Alaimo, K., Packnett, E., Miles, R. A., Kruger, D. J. (2008). Fruit and vegetable intake among urban community gardeners, *Journal of Nutrition Education and Behavior* 40(2), 94–101. https://doi.org/10.1016/J.JNEB.2006.12.003

Andersson, E., Barthel, S. (2016). Memory carriers and stewardship of metropolitan landscapes, *Ecological Indicators* 70, 606–614. https://doi.org/10.1016/J.ECOLIND.2016.02.030

Andersson, E., Barthel, S., Borgström, S., Colding, J., Elmqvist, T., Folke, C., Gren, Å. (2014). Reconnecting cities to the biosphere: Stewardship of green infrastructure and urban ecosystem services, *AMBIO* 43(4), 445–453. https://doi.org/10.1007/s13280-014-0506-y

Breuste, J. H. (2010). Allotment gardens as part of urban green infrastructure: Actual trends and perspectives in Central Europe. *Urban Biodiversity and Design* 463–475. http://doi.org/10.1002/9781444318654.ch25

Camps-Calvet, M., Langemeyer, J., Calvet-Mir, L., Gómez-Baggethun, E. (2016). Ecosystem services provided by urban gardens in Barcelona, Spain: Insights for policy and planning, *Environmental Science & Policy* 62, 14–23. https://doi.org/10.1016/J.ENVSCI.2016.01.007

Caputo, S., Schwab, E., Tsiambaos, K. (2016). Emergent approaches to urban gardening. In *Urban Allotment Gardens in Europe*. New York: Routledge, pp. 229–253.

Coles, R., Costa, S. (2018). Food growing in the city: Exploring the productive urban landscape as a new paradigm for inclusive approaches to the design and planning of future urban open spaces, *Landscape and Urban Planning* 170, 1–5. https://doi.org/10.1016/J.LANDURBPLAN.2017.10.003

Costa, S., Fox-Kämper, R., Good, R., Sentić, I. (2016). The position of urban allotment gardens within the urban fabric. In Bell, S., Fox-Kämper, R., Keshavarz, N., Benson, M., Caputo, S., Noori, S., Voigt, A. (eds). *Urban Allotment Gardens in Europe*. Routledge, pp. 221–228.

Csikszentmihalyi, M., Abuhamdeh, S., Nakamura, J. (2005). Flow. In Elliot, A. J., Dweck, C. S. (eds). *Handbook of Competence and Motivation*. London: Guilford Publications, pp. 598–608.

Delshammar, T., Partalidou, M., Evans, R. (2016). Building trust and social skills in urban allotment gardens. In Bell, A. V. S., Fox-Kämper, R., Keshavarz, N., Benson, M., Caputo, S., Noori, S. (ed). *Urban Allotment Gardens in Europe*. Routledge, pp. 342–363.

Dubbeling, M. (2011). Integrating urban agriculture in the urban landscape, *Urban Agriculture Magazine* 25, 43–46.

Hardman, M., Chipungu, L., Magidimisha, H., Larkham, P. J., Scott, A. J., Armitage, R. P. (2018). Guerrilla gardening and green activism: Rethinking the informal urban growing movement, *Landscape and Urban Planning* 170, 6–14. https://doi.org/10.1016/J.LANDURBPLAN.2017.08.015

Hensley, N. (2015). Cultivating biophilia: Utilizing direct experience to promote environmental sustainability, *Journal of Sustainability Education* 9, 2151–7452.

Hopkins, R. (2019). *From What is to What if: Unleashing the Power of Imagination to Create the Future We Want*. Hartford: Chelsea Green Publishing.

Ioannou, B., Morán, N., Sondermann, M. Certomá, C., Harman, M. (2016). Grassroots gardening movements. In Bell, S., Fox-Kämper, R., Keshavarz, N., Benson, M., Caputo, S., Noori, S., Voight, A. (eds). *Urban allotment gardens in Europe*. New York: Routledge, pp. 62–90.

Kingsley, J. Y., Townsend, M., Henderson-Wilson, C. (2009). Cultivating health and well-being: Members' perceptions of the health benefits of a port Melbourne community garden, *Leisure Studies* 28(2), 207–219.

Kurtz, H. (2001). Differentiating multiple meanings of garden and community, *Urban Geography* 22(7), 656–670. https://doi.org/10.2747/0272-3638.22.7.656
Langemeyer, J., Camps-Calvet, M., Calvet-Mir, L., Barthel, S., Gómez-Baggethun, E. (2018). Stewardship of urban ecosystem services: Understanding the value(s) of urban gardens in Barcelona, *Landscape and Urban Planning* 170, 79–89. https://doi.org/10.1016/J.LANDURBPLAN.2017.09.013
Litt, J. S., Soobader, M. J., Turbin, M. S., Hale, J. W., Buchenau, M., Marshall, J. A. (2011). The influence of social involvement, neighborhood aesthetics, and community garden participation on fruit and vegetable consumption, *American Journal of Public Health* 101(8), 1466–1473. https://doi.org/10.2105/AJPH.2010.300111
McGuire, L., Morris, S. L., Pollard, T. M. (2022). Community gardening and wellbeing: The understandings of organizers and their implications for gardening for health, *Health & Place* 75, 102773. https://doi.org/10.1016/J.HEALTHPLACE.2022.102773
Nakamura, J., Csikszentmihalyi, M. (2009). Flow theory and research. In Snyder, C. R., Lopez. S. J. (eds). *Oxford Handbook of Positive Psychology*. Oxford: Oxford University Press, pp. 195–206.
Noori, S., Benson, M. (2016). Urban allotment garden. A case for place-making. In Bell, S., Fox-Kämper, R., Keshavarz, N., Benson, M., Caputo, S., Noori, S., Voight, A. (eds) *Urban Allotment Gardens in Europe*. New York: Routledge, pp. 291–319.
Schunko, C., Brandner, A. (2022). Urban nature at the fingertips: Investigating wild food foraging to enable nature interactions of urban dwellers, *Ambio* 51(5), 1168–1178. https://doi.org/10.1007/s13280-021-01648-1
Scott, A., Dean, A., Barry, V., Kotter, R. (2018). Places of urban disorder? Exposing the hidden nature and values of an English private urban allotment landscape, *Landscape and Urban Planning* 169, 185–219
Speak, A. F., Mizgajski, A., Borysiak, J. (2015). Allotment gardens and parks: Provision of ecosystem services with an emphasis on biodiversity, *Urban Forestry & Urban Greening* 14(4), 772–781. http://doi.org/10.1016/j.ufug.2015.07.007
Steel, C. (2020). *Sitopia: How Food Can Save the World*. London: Chatto & Windus.
Tilley, C. (2006). The sensory dimensions of gardening, *The Senses and Society* 1(3), 311–330. https://doi.org/10.2752/174589206778476252
Viljoen, A. (2005). *CPULs: Continuous Productive Urban Landscapes: Designing Urban Agriculture for Sustainable Cities*. New York: Architectural Press.
Viljoen, A., Bohn, K. (2009). Continuous productive urban landscapes (CPUL): Essential infrastructure and edible ornament, *Open House International* 34(2). www.researchgate.net/publication/44960513
Walter, P. (2013). Theorising community gardens as pedagogical sites in the food movement, *Environmental Education Research* 19(4), 521–539. http://doi.org/10.1080/13504622.2012.709824
Wilson, B. (2022). To eat in the modern world is often to eat in a state of profound sensory disengagement. It shouldn't have to be this way, *The Guardian*. www.theguardian.com/food/2022/mar/29/how-we-lost-our-sensory-connection-with-food-and-how-to-restore-it
Wilson, E. O. (1984). *Biophilia*. Cambridge: Harvard University Press.
Wood, C. J., Pretty, J., Griffin, M. (2016). A case – Control study of the health and well-being benefits of allotment gardening, *Journal of Public Health* 38(3), e336–e344. https://doi.org/10.1093/PUBMED/FDV146

8

MOVING FROM THEORY TO PRACTICE

Approaches and Parameters

Introduction

In earlier chapters, questions were raised concerning how we dwell in nature as an urban society, with the related aspect of whether current approaches to urban landscapes are fit for purpose. Our relationship with nature and the natural world is increasingly being defined by urban lifestyles and associated activities, which involve a modern interpretation of nature and the landscape. This has to address the range of urban environments that are present or being created, and how a high-quality environment should be defined as such, with the imperative that we have to respond appropriately in the light of current knowledge and concerns. The corollary is that the ways that we interface with the environment are still being defined by our inherent need to experience nature, where it is clear to see that its presence acts as a positive pathway to health and well-being and that the absence of contact with the natural world, seen in many urban development models, has failed to provide the biophilic stimulation that is so easily accessed. In the urban world, it is not just that nature has to be reintroduced, but that to a certain extent it needs to be (and is being) re-interpreted in respect of the opportunities that exist to embrace its power throughout the urban realm as a universal quality that underpins urban living at all levels. How we interface with natural or close-to-natural systems in our daily lives is thus a fundamental consideration which has become more evident as we have become withdrawn from this natural relationship.

The connections associated with accessing the natural world, involving positive mental and physical states of health, are sufficiently well articulated and of sufficient importance to consider what an urban biophilically focused landscape might look like and the processes by which it is to be achieved. There is now plenty of evidence and examples to demonstrate a clear desired trajectory regarding the qualities of the urban environment in relation to health and well-being and the

DOI: 10.4324/9781003099758-9

underpinning importance of the human-nature bond indicated in the biophilia hypothesis. Accordingly, in this chapter, we address this situation to move from theory to practice by considering some of the key aspects which are defining the arena of practice, and in the follow-on chapter (Chapter 9), to present a range of case studies which address different approaches and situations. In selecting case studies, while high-profile international studies are impressive and important for showing what can be done, we have opted for a balanced presentation of these and local studies which have particular relevance to everyday situations (yet are rarely featured), including those that are specifically related to the authors' experience. These are analysed to demonstrate their respective qualities and the various issues that they address, and here we focus on some of the key themes arising from earlier discussion. Accordingly, this chapter is, essentially, intended as a primer to help make the reading of these case studies somewhat easier, and so that they can be appreciated in the totality of exploration, approaches and opportunities.

Adopting an Ecocentric Approach – the Socio-Ecological System

The associated parameters that relate to investigating and understanding how we interact and are influenced by the landscape are best understood by taking an eco-centric approach. This is an approach that considers all aspects of the environment under the remit of ecological associations, and where the various facets of the urban environment can and should be explored in relation to human-based ecological dimensions which embrace nature conservation as an important part of its remit. Acknowledging this, it is pertinent to use the hybrid term 'socio-ecology' to describe the system (Frank et al., 2017; Petrosillo et al., 2015).

Socio-ecology embraces the importance of the basic human-nature bond (for example, see Kahn & Kellert, 2002), restoring and supporting this bond and understanding the dynamics of person/place interaction, especially in relation to biophilic thinking. It is also indicated by studies which link social activities, place identity and accessing nature as a positive combination of factors that support communities (Petrosillo et al., 2015). These address issues of how the natural landscape is perceived and used, where there is a range of theoretical perspectives that explain ideas and give consistent messages about the importance of biophilia, but also concern the specifics and diversity of the urban landscape. This includes how the city is read (Chapter 1), how connections are made, or indeed impeded, by different parameters, including how the senses respond, the pace rhythm and flows of walking and how we associate with spaces (Chapter 1). By embracing theory and analysing practice, such ideas tell us how people move through and interact with different environments in ways that stimulate us as individuals to support positive states of mind and health. Indeed, in presenting ideas in the earlier chapters, the dynamics of such a socio-ecological system are already substantially explored. However, they also present a certain amount of challenge regarding what constitutes the public realm, especially concerning the wider aspect of how we view the landscape and the responsibilities of all in assuring that the features of these socio-ecological systems function together in an effective way. In this respect, the socio-ecological concept is a particularly useful one in that it does not recognise artificial boundaries of ownership, although it does recognise natural boundaries; indeed, it is these types of boundaries which give places identity and feature in how environments are read or understood. This understanding is well demonstrated in biophilic cities, where biophilia is seen as the driver of urban form in which all other aspects can be considered, essentially expressing the quality of different socio-ecological city systems (biophiliccities.org; Russo & Cirella, 2017).

Current thinking thus moves towards an understanding of 'urban' as an ecosystem in its own right. This is a system which embraces a complex of sub-systems which are understood from a wider environmental perspective, including their performance, resilience and sustainability, and which need to be maintained or developed in ways that environmental quality is not compromised – in reality, it needs to be substantially improved to maintain conditions that are suitable for high-quality 21st-century living and to address wider environmental concerns.

In theory, as far as a socio-ecological relationship is involved, it should be that the combination of the natural, designed and built forms (Almusaed, 2006) found in urban areas provide an increased number of socio-ecological niches, as well as containing wilder places set aside more specifically for nature, all of which can be explored and understood in terms of their environmental performance. While it is proper and essential that places be considered in relation to their wildlife value and are appropriately protected, as wild places, they also form a critical cornerstone of a socio-ecological system. The human-nature bond, biophilia and natural processes underpin the concept of socio-ecology, but the qualities of natural systems would seem to provide an appropriate base for considering urban ecosystems in relation to urban dynamics and is becoming more fundamental to thinking, particularly in the concepts of ecosystem services and biophilic design (Bennet & Chaplin Kramer, 2016; Kellert & Calabrese, 2015).

Responsibilities and Remits

These are ideas that can be addressed from different perspectives: professional, personal and community. For example, in terms of professional practice, there are disciplines which have a particular responsibility in designing, planning and managing our environment, especially environmental planning, architecture, landscape architecture and urban design. Activities within these sectors do demonstrate ideas and design solutions which can be considered inspirational in their thinking, which need to be highlighted in showing possible ways forward and which, on examination, apply the very principles that appear from investigation. In reality of course, it is the totality of many disciplines across the sciences, arts, humanities and medicines, including health provision, social sciences and education, which are impacted by current thinking, but are also responding to the current environmental agenda. These disciplines explore solutions within the context of their professional remits, but in reviewing ideas, it is clear that they can be embraced within socio-ecological thinking where they actually define and articulate how the system functions in multiple ways. While not necessarily starting from the perspective of environment or nature, they arrive at the same socio-ecological destination. However, this is only half the story, since it needs an approach where the views and desires of local people are fully understood, accommodated and given appropriate emphasis. This involves several considerations:

- That urban ecosystem functioning is understood and responsibilities and roles readily accepted, including personal responsibilities.
- That people have the knowledge, ability and skills to make best use of their local environment in respect of their own quality of life.
- That potential pathways and measures are recognised, understood and applied.

Realising and Releasing the Power of Nature

Discussion thus moves consideration towards understanding and maintaining wider ecosystem performance as the underlying factor in terms of constituting high-quality environments for human occupation and a definable contribution to the delivery of sustainable urban infrastructure (Project Learning Tree, n.d.). This does appear to be the current trajectory, but it still presents some challenges to existing norms, including the ways in which environmental quality is evaluated, the priority that it is given and what parameters are used to define it. However, by taking an ecocentric approach, discussion is moved to consider: what is the basic relationship that exists between people and their environments, and how is this relationship understood, measured and re-enforced in practice?

These ideas begin to explain the complexities and integration of green infrastructure as a system which embraces the entire urban scene as a complex of sub ecosystems. It also sets the primary aspect of enquiry and performance to that concerning the impact upon populations and individuals, set firmly within the wider debate regarding the city as a legitimate ecosystem which functions and is required to function within sustainable parameters (Tebb, 2020). If the ecological connections, flows and impacts of the environment are understood, then it is possible to develop and manage urban areas, including retrofitting, according to informed environmental strategies. Our current state of knowledge shows that these parameters are already reflected in the various facets of investigation, clearly concerning health and well-being, but also in recommendations and minimum standards: for example, concerning the standards of access and spatial distribution of green spaces across the urban scene recommended by WHO and others (Chapter 3). By taking an ecocentric stance, managing and developing the urban environment moves from a process that tries to satisfy many conflicting priorities, to one where the environment contains and delivers different outcomes as an integrated whole. It thus moves consideration to one where different urban sub-ecosystems are considered as part of an overall complex, rather than being treated as unconnected and bounded components, which in reality is not the case, understanding how the linkages play out with an associated reappraisal of performance.

In this respect, the ideas discussed in the previous chapters give a fairly diverse exploration of human environment relationships with emphasis on the human-nature bond, as well as addressing a range of urban environments from natural to designed, which can and should be considered as part of an interconnected socio-ecological complex. As presented in the earlier chapters, these investigations are across disciplines which address different urban environments and circumstances. In other words, there is enough information out there to move forward and adopt an ecocentric approach in assessing and defining urban quality, as well as urban splendour, through the recognition of urban situations that show off a city's biophilic credentials and make impressive statements regarding their performance (biophiliccities.org; Beatley, 2011; 2009).

The move to an ecocentric approach is in fact already well established concerning the international adoption of ecosystems services (Bennet & Chaplin Kramer, 2016), addressing concerns regarding environmental degradation, climate change and species extinction, as well as ideas relating to re-wilding (Sturgeon, 2021), the restoration of habitat and creation of new biophilically focused landscapes, such as urban agriculture, and built forms as ecosystems in their own right (Almusaed, 2006). These apply existing knowledge and thinking and show that change is

FIGURE 8.1 The invitation of nature and the dilemma of urban, Cripplegate Park: good-quality biophilic places exist throughout the urban scene, but are typically weakened by the urban infrastructure of the modern traffic-congested city, a situation that requires resolution.

coming and necessary concerning the ways that we understand, articulate and assess the qualities and performance of urban. As suggested, it does challenge accepted norms, but also offers exciting possibilities, especially within the field of biophilic design and biophilic cities, where an ecocentric approach is the basis of professional activities and city identity (Tebb, 2020). It also underpins the reactions of many to embrace (or restore) nature as important aspects of their everyday lives (Mumaw & Raymond, 2021); to seek out opportunities to realise this ambition in the ways that they look after their own personal spaces or visit landscapes that offer the same qualities as a respite from the stresses of urban living, or perhaps offer perfect venues for family or community events (Coles & Caserio, 2015). That such opportunities are not offered routinely in every part of the urban scene when the parameters are clear seems somewhat perverse in the light of current knowledge.

Environmental Generational Amnesia

One of the cultural challenges to upholding the human-nature bond and realising biophilic satisfaction is a phenomenon known as 'environmental generational amnesia', identified by Kahn (2002). Kahn discusses how the baseline for what is considered a normal environmental condition shifts as it continues to degrade. As environmental degradation continues, the baseline continues to shift with each ensuing generation, each perceiving this degraded condition as the norm or non-degraded condition (Kahn, 2002).

Kahn's idea of environmental amnesia goes some way to explaining how everyday contact with nature has continued to diminish, i.e. what is considered as a normal condition and thus acceptable is in reality a degraded one. Accordingly, there are subsequent low expectations concerning environmental quality which obscure the reality of the retreating presence of the natural world, including accessibility to it, especially in respect of people's everyday lives. Similarly, it is suggested that the process also serves to foster a division between the urban populated/dominated world, where nature is not well represented, with wilder places where it is, as an accepted norm. Thus, there has been a consequent failure to fully embrace what could be. This is a situation which, in the light of increasing urbanisation, is clearly unsustainable and is likely to continue unless the cycle of amnesia is broken and that associated cultural barriers are challenged. Re-establishing connections with nature and supporting people, especially children, to understand and experience nature firsthand is critical to breaking this cycle. In parallel, identifying examples of what the urban environment could be serves to raise expectations and challenge norms, and involves a suite of measures that both stimulate the biophilic self and re-establish direct contact with high-quality ecosystems (Box 8.1).

An understanding of what a healthy home environment should look like, i.e. when it demonstrates the parameters of a healthy intact ecosystem, needs appropriate reference points, including precedents and exemplars to show how rich such urban ecosystems can be. These are of course the visions which underpin biophilic design and the international biophilic cities movement, but must also be looked for regarding the more routine parts of our urban centres (Tebb, 2020).

BOX 8.1 CHALLENGING NORMS

- Adopt an ecocentric stance in order to understand the structure and functioning or the urban environmental within the specific context of human-environment relationships and the human-nature bond.
- Move towards a fuller understanding of what healthy and intact urban ecosystems could look like. Identify exemplars of urban forms which demonstrate the dynamics of urban environments and publicise how rich such ecosystems can be.
- Break the cycle of environmental generational amnesia and raise environmental ambitions in professional and public arenas, including education, to a high level that equates with current knowledge.
- Encourage a stewardship approach which embraces all stakeholders and thus engenders a relationship which is sustainable.
- Apply current ideas and thinking concerning environmental interaction and biophilia to develop the required socio-ecological linkages throughout the urban scene.
- Adopt existing recommendations and standards with particular reference to local centres of population.

Defining the Arena of Practice

Discussion and studies, especially in the field of environmental psychology (and reviewed in Chapters 1 and 2), which focus on the restorative power of nature are relatively easily explained by the biophilia hypothesis, and point to the importance of wild places that are easily and freely accessible both within and close to urban centres. Similarly, places which have landscape properties that approximate nature-like conditions, a condition that might be called biophilically saturated, offer similarly related opportunities. However, more detailed enquiry gives a clear indication of how the balance between urban influences and biophilic influences is played out as two opposing forces, which result in the experience being a positive one, fostering a desire to explore it further, or negative, with a desire to move away (Chapter 4).

What is clear from these studies is that people, given the appropriate confidence, naturally align themselves with the properties of natural environments and seek out positive biophilic experiences across a range of situations, from moments of chilling out to deeper immersive experiences. Such findings endorse the adoption of a biophilic approach across the spectrum of urban situations and the importance of satisfying the human-nature bond. Thus indicated is a relationship which concerns the restorative impact of nature and the generality of living within a nature/biophilic setting as part of an everyday pattern, plus easily accessible and appropriate 'special' environments which satisfy a desire for many to seek out moments involving deeper experiences as part of their personal life strategies. Key aspects of such include the opportunity to get away from the everyday, but also to create a

FIGURE 8.2 Eastside Park Birmingham: a new park for the city centre offering a range of sensory affordances, including the sound of water, light and shade, places to sit and walk and robust planting.

positive urban ambience through the development of biophilically focused green infrastructure, ensuring that many opportunities and locations exist throughout the urban realm and are of a scale to support varied activities (Figures 8.2 and 8.3).

Indeed, this is a position postulated in the introduction to Chapter 2 (concerning the study of the housing area of Castle Vale), which sees the biophilically focused infrastructure as a routinely accessible resource that can be experienced/ used by everyone on whatever basis they wish, but also to which they are continuously exposed, and that is capable of providing a restorative experience. In the original study, ideas were compared to the process of homeostasis, the mechanism by which the body keeps itself in a steady state by adjusting its internal metabolic rhythms, but extending this to include the positive influence of and exposure to a biophilic environment as a type of preventative or prophylactic measure to maintain balanced health. The various studies, relationships and theories (i.e. those presented throughout the chapters, which articulate the experiences of individuals) suggest that such an analogy is apt and seems to underpin current thinking regarding accessing nature as a health-related activity, seeking out biophilic and nature experience as a mechanism to help maintain or restore an appropriate mental and physical balance. The indicated direction for urban landscapes is thus relatively clear: to restore connections by retrofitting appropriate green infrastructure to ensure that these are met and in ensuring that wilder places are well represented within or close to urban centres,

Biophilic situations appear to stimulate the body, which is naturally aligned to nature via the human-nature bond. They have properties where they offer a restorative experience related to well-being and 'getting away', but also promote further exploration, which involves eco-haptic behaviours: a sense of recreational visual and sensory foraging, involving moderate physical activity (Chapter 5). Nature

FIGURE 8.3 Dwelling in nature: new wildlife ponds at University of Birmingham campus.

settings seems to be particularly satisfying in terms of the connections, flows and feedback that are involved. In practice, but assuming that people have to confidence to access them, nature sites do this automatically, but clearly such attributes can be and should be designed into urban places and forms the mainstay of biophilic design. Quality biophilic environments can thus be defined by their visual and eco-haptic qualities, the degree to which they offer separation from the everyday, the level of sensory stimulation offered through haptic and visual connections and also the extent to which they are accessible. It is in this context that recommendations are made regarding green infrastructure typologies and the minimum standards regarding the location of extensive green spaces (Chapter 3). They also define the modern context of the experience as one of leisure, recreation-related or restful activities, including the idea of being away from the everyday, and indicate the direction for an urban landscape aesthetic which can be derived from such understanding.

Person-Place Relationships and the Socio-Eco – the Context of Environmental Perception

In the studies reviewed, authors emphasise that how people perceive the environment is highly personal. While there may be common influences, an encounter is given specific meaning by how it is contained, formulated and rationalised as a personal construct. This is true across the range of environments and situations explored in earlier chapters. For example, in recalling events surrounding an encounter, the stories told relate to such aspects as family events or past recollections where the current experience is interpreted in the light of previous experiences (Chapter 4). That this involves both cultural and nature focused processes is inherent in the system. In addition, the adoption of an ecocentric approach is evident across studies. For example, Lynch, in his examination of the 'Image of the City' (Chapter 1), describes the spatial organisation of the city, its attributes and how it is defined by ideas of 'nodes, edges, focal points' and views, which can be considered as a subset terminology (expressed in an urban design remit) regarding an urban socio-ecological system. While Lynch does not make reference to this term, or to biophilia, he does make reference to places that are clearly biophilic. Gehl (Chapter 1) also adopts an urban design perspective in considering the public realm, and essentially takes an ecocentric and socio-ecological interpretation concerning how the body orientates itself in relation to urban space in ways that are positive, how these aspects start to define the quality of the public realm and underpin design thinking.

Taking an ecocentric stance, considered in conjunction with the dynamics associated with moving through and experiencing nature, and the dynamics described in relation to a city's structure, points towards a fuller understanding of person-place relationships, be they nature-focused or urban-focused. Ideas have substantial commonality, and together it is suggested that they underpin consideration of urban aesthetics set within the ecocentric focus of socio-ecology, which falls under the remit of urban, especially design-based, professions. This involves an aesthetic which has been informed by the reality of people's experiences and reactions, and clearly can and should be contained within local initiatives as part of a broader notion of environmental stewardship (Mumaw & Raymond, 2021). One only needs to refer back to the earlier studies presented which address different urban situations, and how the associated spaces were used to see people-place dynamics at work, including the haptic, visual and emotional connections associated. However, one does have to take into account the problem of environmental amnesia as a barrier and how this barrier can be removed, which has meant that isolation from nature has caused a general reduction in confidence in understanding and embracing our biophilic roots (Kahn, 2002).

Given the wide definition of urban nature suggested in Chapter 1, and similarly as embraced in biophilic design (Chapter 2), it is not particularly difficult to visualise the urban environment as offering a range of biophilic delights (Cabanek et al., 2020): views, walking through wilder areas, parks, avenues and clusters of street trees, opportunities and invitations to stop, sit and chill out, as well as containing those nature places that have special meaning, but which also draw on and extend understanding of how the city is read and how public space is organised. In addition, studies suggest that adopting a personal approach which tends towards the mindful (Chapter 5) is a positive move in increasing an awareness of the delights of nature and its therapeutic impact. One only has to consider the conscious preference to walk through a local park rather than take the more direct route alongside traffic to realise that these ideas are simply a re-enforcement and expansion of what is already a natural and familiar process, but also to understand the draw of such locations and the sensory affordances they offer.

Pathways

It is clear that the landscape acts as a moderating factor in an urban world where people respond positively to natural features, and indicates that there is a direct relationship between the biophilic qualities of places and the extent to which they are considered attractive and 'good places to be'. This is the relationship explored in biophilic design, but also points towards retrofitting to improve the biophilic properties of places. Similarly, the same thinking applies regarding expanding opportunities to encounter these types of quality places, including how they are represented in 'the view'. As a mechanism for gaining greater satisfaction from the environment, biophilic interventions thus form a pathway for health and well-being and suggest a relatively simple process of intervention, but at the same time remove barriers that prevent the system operating in its most effective way (Totaforti, 2020). As a pathway, it is necessary to look at urban forms, places new and existing, through a biophilic lens and not be put off by issues of ownership or existing restrictions, but to seek inspiration from examples and others regarding what can be done and what could be, understanding that each situation represents a particular opportunity for action in the knowledge that each forms part of the wider socio-ecological system.

In terms of the specifics surrounding biophilia and nature, the ideas reviewed suggest that, in relation to environmental engagement, biophilic connections are a key driver and that they produce a response that is satisfying in the way that the senses are stimulated, and that the experience is an embodied one and correspondingly emotionally satisfying. In terms of practice, this means that attention should be paid to both the view, place qualities and the sequencing of spaces in terms of the ways that they offer such properties, textures, touch, sounds, refuge, etc., including how the speed of walking might be moderated, views offered, or indeed explained and highlighted. This includes specific initiatives relating to physical activity, in that these are environments which appear inviting, excite the senses and compel people to investigate them further and encourage sustained physical activity. It would seem that the human-nature bond stimulates both relaxation and action, achieving biophilic satisfaction in sensory and physical ways, felt through natural and satisfying physical achievement and emotional contentment.

However, it is also necessary to provide support mechanisms to ensure that individuals have the capacity and confidence to comfortably access places which offer such opportunities. Clearly, a range of complementary activities aid this process; the role of the arts was discussed in Chapter 5, as was the role of contemporary web-based social platforms. Combining the two together would suggest a sophisticated suite of initiatives that would underpin biophilic interventions utilising both the power of the arts and the power of nature, plus the ways that they are able to

FIGURE 8.4 Getting wilder. Open spaces remain a vital part of nearby nature, and they are getting wilder as nature is allowed to reclaim its territory.

Source: Image credits: Susan Coles

challenge norms and celebrate environmental achievement, but also help to give direction in promoting a mindful approach, or indicating permission to engage with the environment. Accordingly, while direct action to improve the properties of places is absolutely key, this has to be set within a wider initiative of education and support, celebration, communication and collaboration. The socio-ecological system is thus defined by both the quality of its constituent places, but also the niches of activity which concern realising opportunities for different types of action and addressing different sectors, private, public, community, individual and virtual, to best effect. This would of course also demonstrate the quality and sophistication of the system. Each biophilic system would thus be defined by the quality of its environment measured in both nature conservation, education and social terms and the extent to which the system is supported by actions and understanding applied at all scales (biophiliccities.org).

While there is a ways to go in achieving this, this is the trajectory indicated, and while some initiatives might be rather isolated at the moment, considered together they do demonstrate what could be done and what the urban future and urban living could be. It is in this context that the case studies have been selected to demonstrate the different pathways and niches that are being explored and the enthusiasm by which they are being pursued, which in itself might be an indicator of the satisfaction that can be derived in exploring biophilic solutions.

Conclusions

This chapter has reviewed a range of ideas discussed in earlier chapter, under the remit of moving from theory to practice, and is intended to serve as a primer to the case studies to be considered in the follow-on chapter. These ideas recommend adopting an ecocentric approach and considering the urban environment as a socio-ecological system which is focused on its biophilic attributes. The importance of the human-nature bond is emphasised, and ideas are moving to consider how this relationship might be underpinned in practice.

This discussion concerned the general remit of professionals, but also emphasised that people need the confidence and skills to achieve satisfaction from accessing biophilic related experiences. In this context, the issue of Generational Environmental Amnesia is raised regarding the acceptance of a deteriorating environment, and that a range of interventions are likely to be required to break this cycle.

Reference is made to pathways to actions including achieving biophilic satisfaction, while the socio-ecological system is also considered as offering quality environments, but also embracing the initiatives and activities which underpin confident use and promote understanding of the environment. In this context, reference is made to biophilic cities, which can be seen to demonstrate these qualities and to which other cities might aspire.

References

Almusaed, A. (2006). Biophilic architecture: Towards a New Potential of Healthy Architecture (Conference: Rethinking Sustainable Construction), *12th Rinker International Conference, Next Generation Green Buildings, at Sarasota, Florida*, Vol. 1. researchgate.net

Beatley, T. (2009). Biophilic urbanism: Inviting nature back into our communities and into our lives, *William & Mary Environmental Law and Policy Review* 34(1), 6. https://scholarship.law.wm.edu/wmelpr/vol34/iss1/6.

Beatley, T. (2011). *Biophilic Cities: Integrating Nature into Urban Design and Planning*. Washington, DC: Island Press.
Bennett, E. M., Chaplin-Kramer, R. (2016). Science for the sustainable use of ecosystem services, *F1000 Research* 5. https://doi.org/10.12688/F1000RESEARCH.9470.1
Cabanek, A., Zingoni de Baro, M. E., Newman, P. (2020). Biophilic streets: A design framework for creating multiple urban benefits, *Sustainable Earth* 3(1), 1–17. https://doi.org/10.1186/S42055-020-00027-0
Coles, R. W., Caserio, M. (2015). *Social Criteria for the Evaluation and Development of Urban Green Spaces* (Report number D7 EVK4-CT-2000–0002). www.researchgate.net/publication/288181939_Social_Criteria_for_the_Evaluation_and_Development_of_Urban_Green_Spaces
Frank, B., Delano, D., Caniglia, S. (2017). Urban systems: A socio – Ecological system perspective, *Sociology International Journal* 1(1), 1–8. https://doi org/10.15406/sij.2017.01.00001
Kahn, P. H. (2002). Children's affiliations with nature: Structure, development, and the problem of environmental generational amnesia. In Kahn, P. H., Kellert, S. R. (eds). *Children and Nature: Psychological, Sociocultural, and Evolutionary Investigations*. Cambridge: MIT Press.
Kahn, P. H., Kellert, S. R. (2002). *Children and Nature: Psychological, Sociocultural, and Evolutionary Investigations*. Cambridge: MIT Press.
Kellert, S., Calabrese, E. (2015). *The Practice of Biophilic Design*. www.biophilic-design.com
Mumaw, L. M., Raymond, C. M. (2021). A framework for catalyzing the rapid scaling of urban biodiversity stewardship programs, *Journal of Environmental Management* 292, 112745. https://doi.org/10.1016/J.JENVMAN.2021.112745
Petrosillo, I., Aretano, R., Zurlini, G. (2015). Socioecological systems. In Elias, S. A. (ed). *Reference Module in Earth Systems and Environmental Sciences*. Amsterdam: Elsevier Publisher, pp. 1–7.
Project Learning Tree. (n.d.). *Innovative Ways to Create more Urban Green Spaces*. Project Learning Tree. plt.org.
Russo, A., Cirella, G. T. (2017). Biophilic cities: Planning for sustainable and smart urban environments. In Ajaz, R. (ed). *Smart Cities Movement in Brics*, pp. 153–159. www.globalpolicyjournal.com/projects/gp-e-books/smart-cities-movement-brics
Sturgeon, A. (2021). Rewilding our cities: Beauty, biodiversity and the biophilic cities movement, *The Guardian Newspaper*, 4 April. www.theguardian.com/artanddesign/2021/apr/05/re-wilding-our-cities-beauty-biodiversity-and-the-biophilic-cities-movement
Tebb, P. J. (2020). *Biophilic Urbanism: Designing Resilient Communities for the Future*. Abingdon: Routledge.
Totaforti, S. (2020). Emerging biophilic urbanism: The value of the human-nature relationship in the urban space, *Sustainability* 12(13), 5487. https://doi.org/10.3390/su12135487

9

CASE STUDIES

9.1 Case Study 1 – Arup Cities Alive – A Green-Infrastructure-Led Design Approach

In this case study, we look at Arup's Cities Alive initiative, which involves an exploration of the role of green infrastructure (GI) in building urban-based ecosystems that show high performance in terms of their socio-ecological functioning. The emphasis is on 'a green-infrastructure-led design approach', developing strategies and moving to a situation where the various GI niches are understood and fully exploited to help realise the ambition of the sustainable city, which also accords with biophilic principles and how ideas might be applied to the advantage of nature, environment, economy, health and society.

> Cities Alive proposes a design approach for urban environments that promotes nature as a key driver . . . that acknowledges the essential value of the natural environment and its role in underpinning our economic prosperity, health and wellbeing. The approach seeks to create healthier more socially cohesive and biodiverse urban environments and a connected city ecosystem for people and wildlife.
>
> *(Cities Alive: rethinking green infrastructure)*

Cities Alive was developed by Arup's Foresight, Research, and Innovation Team, which

DOI: 10.4324/9781003099758-10

> identifies and monitors the trends and issues likely to have a significant impact upon the built environment and society at large, researching and raising awareness about the major challenges affecting the built environment and their implications.
>
> *(Cities Alive website)*

Two publications in the series are targeted here as being particularly relevant to discussion: 'Cities Alive: Green Building Envelope' and 'Cities Alive: Rethinking Green Infrastructure'. Both are available as downloads from the Arup, Cities Alive website, and address city futures and the approaches indicated in understanding and developing sophisticated green infrastructure systems. Ideas are presented here through a series of images which demonstrate different aspects of practice, plus selected extracts taken from these two publications. Where appropriate, additional explanatory commentary links examples to ideas explored in earlier chapters.

- Cities Alive expounds the significant contribution that a GI-led design approach can make towards addressing social, environmental and economic issues in our urban environments, whilst also contributing to climate change resilience.
- Cities Alive shows how the creation of a linked city ecosystem encompassing parks and open spaces; urban trees, streets, squares; woodland and waterways can help create healthier, safer and more prosperous cities. How we can build nature into our urban systems at all scales through high quality landscape design, via new development or retrofitting through a green infrastructure design approach.
- To realise this vision, green infrastructure has to now take a more influential role in the planning and design of cities and urban environments.

(See Arup, Cities Alive: Green Infrastructure)

In their exploration and vision of GI, Cities Alive essentially takes an eco-centric approach, which recognises that all elements of the city are linked and constitute parts of a wider connected system. In ecological terms, this is a system where artificial boundaries of ownership are of no actual consequence and thus considers the inclusivity of the entire urban scene, urban spaces and built forms as proper targets for improving green infrastructure, be it in conventional spaces or, for example, as vertical elements which clad or top buildings. Ideas progress the GI agenda on a number of levels: in demonstrating precedents and visions of 'what could be', and moving to a situation where urban forms start to express their full potential as sub-ecosystems but operating within a whole. These address the universality of ecocentric thinking and application across different urban situations, including the need to operate at all scales. Essentially, they encompass the types of ideas that will move 'urban' towards a healthy ecological condition, which shows resilience and sophisticated levels of environmental performance underpinned by the power of nature (Figure 9.1.1).

> Increasingly, GI is being seen as a concept which unites a range of disciplines and interests, and that facilitates collaborative working. Crucially, in the context of green infrastructure, competing priorities can often complement each other.
>
> *(Arup, Cities Alive)*

Collaboration is emphasised, in which:

- 'Strengths, priorities, opportunities, and requirements of different actors must be considered in order to acknowledge the political nature of delivery, in particular across local boundaries.'
- 'Local authorities, developers, clients, landowners, utility providers, the community and the built environment professionals should communicate, share knowledge and educate others in the benefits of GI.'

FIGURE 9.1.1 Green Building Envelope: balancing built and natural forms.

Source: Image credits: Daniel Imade; Isabel Heinmann; Arup

- 'All actors should contribute to the vision for GI, with the aim of identifying interventions that are able to adapt to changing contexts and the needs of the different actors involved.'
- 'Planners should always negotiate, allowing new opportunities to be delivered as they emerge. Crucial to any negotiation is the ability to promote the case for GI to those responsible for delivery in a way that is appropriate to their needs.'

(Arup Cities Alive: Rethinking Green Infrastructure)

The GI Ecosystem

Cities Alive emphasises that the green infrastructure system needs to be robust, of a sufficient scale to balance built forms and be resilient to the relatively short-term changes of urban re-development. As a socio-ecological complex, the identity and response to the city is influenced by its built forms, nature places and biophilic properties. These define its grandeur and make up the 'image of the city', especially the quality of the public realm (Figures 9.1.2 and 9.1.3). Cities Alive, in promoting a linked 'city ecosystem' and building nature into urban systems at all scales, invites others to examine ideas and precedents which re-conceptualise the city as a healthy functioning ecosystem, defined by both buildings and nature being complementary and in harmony. Different elements of GI are demonstrated in embracing the entire complex of urban forms and urban spaces extending to vertical greening, urban farms and sky gardens, which reference functionality and different arenas of practice. In ecological terms, these can be considered as niches, constituent parts of the wider urban-based socio-ecological system which performs as a whole, i.e. the urban GI forms the backbone of the urban ecosystem, each niche providing identified outputs to support city functioning, population health and the wider provision of ecosystem services.

How the different elements are perceived and utilised and the reality of practice will reflect how people reside in the city, read it and use or interact with the different situations. In such a system, sophistication is provided by the diversity of nature and green forms, aided by urban design principles, including biophilic design, much advantage being taken of retrofitting to fill in the missing gaps and improve the environmental performance of built forms and places. This is an important part of the urban GI story, but the urban realm has to be a balanced one where nature is also present in the form of substantial wilder areas and

FIGURE 9.1.2 St James Park London: public open space defined by scale, grandeur and mature trees: 'Outcomes from research clearly demonstrate that it makes sound financial sense to plant large species of trees in urban areas given their stature and the clear benefits that can be achieved'. Note the scale and quality of the encounter, including the mature trees, the quality of the light and shadows that make this a powerful socio-ecological space. (Cities Alive)

Source: Image credits: Daniel Imade: Arup

FIGURE 9.1.3 Gordon Square, Bloomsbury, London: green space provides a haven for social interaction on a summer's day.

Source: Image credits: Paul Carstairs

related places close to locations where people live and work, i.e. specifically 'set aside'; such has been shown to be a priority in ensuring that such places can offer refuge and immersive experiences, with opportunities for walking and relaxing in nature.

Vertical Greening

As buildings rise, in parallel, the green component is expected to rise and thus maintain the green/built balance. This is increasingly reflected in architectural practice that explores the ideology of the built form as an ecosystem in its own right and the new niches that are offered. For example, the Bosco Verticale, Milan, designed by Stefan Boeri (Figure 9.1.4), represents both a reality of vertical greening and the philosophy behind it, and forms an important precedent and exemplar regarding what is possible or 'could be'. The intention is to re-forest the inner cities with high-rise apartment blocks covered with the maximum amount of vegetation. The Bosco Verticale incorporates more than 900 trees, 5,000 shrubs and 11,000 floral plants, covering 96,000 sq. feet of terraces. (For example, see https://placetech.net/analysis/worlds-smartest-buildings-bosco-verticale).

Critically, vertical greening demonstrates how the concept of built forms as components of the wider ecosystem is evolving, and invites others to consider alternatives to conventional systems at all scales, including green walls and living systems as interventions to create nature-inspired biophilic urban forms which integrate with other approaches (Figure 9.1.5).

FIGURE 9.1.4 Porto Nuova Isola, The Bosco Verticale, Milan, by Stefan Boeri.

Source: Image credits: Arup

FIGURE 9.1.5 Singapore School of the Arts: as buildings rise, new niches are available for developing GI linked to improving building performance.

Source: Image credits: Darren Soh: Arup

New Nature, an Urban Essential

Contact with nature must be provided in the urban scene. Places primarily designed to support high levels of biodiversity are a proper consideration of urban land-use priorities, and this is echoed in a move to adopt a more natural approach to urban landscapes which reflects environmental responsibilities and refined attitudes. In many respects, this can be considered as part of wider re-wilding initiatives to the benefit of both species diversity and the human experience, and opportunities exist to exploit new niches in supporting biodiversity which further expand ideas of what urban ecosystems should be (Figures 9.1.6 and 9.1.7). As an approach, designed vegetation based on natural systems to support biodiversity offers a powerful landscape aesthetic in its own right, as a space to be viewed and as a place for contemplation and relaxation. These may well offer situations where arts installations are included to embed appropriate messages involving landscape/arts collaboration. Looking at, but not necessarily intruding upon, areas like this emphasises the need for regular contact

FIGURE 9.1.6 Queen Elizabeth Olympic Park, Stratford, London: 'A key objective of the development of the Olympic Park in London was to leave a valuable legacy of open space and biodiversity. . . . The hope is to encourage a return of historically native animal species, some of which are now fleetingly rare. This use of the urban environment as opportunity for animal repopulation and preservation will become a crucial part of future planning' (Cities Alive).

Source: Image credits: Thomas Graham: Arup

with the natural world, but also helps to understand the importance of the natural world and how we dwell in nature, including our environmental responsibilities.

Retrofitting and Re-Claiming the City through Nature: The examples of New York High Line and Landschaftspark, Duisberg-Nord

These are well-known and inspirational examples of using nature to repair the damage of industry, providing unique landscapes with their own special characteristics and providing important precedents regarding the recycling of urban land and the re-utilisation of obsolete infrastructure.

The High Line (Figure 9.1.8) is managed by Friends of the High Line, a non-profit organisation under the wider jurisdiction of New York's Parks and Recreation Department. The plant selection has a focus on native species, with over 350 types of perennials, grasses, shrubs, vines and trees. Each year, the Friends of the High Line hosts over 450 programmes

FIGURE 9.1.7 Triton Street Green Roof, London.

Source: Image credits: Arup

and activities while community engagement is encouraged. It has over 20 million visitors each year (see www.the highline.org/about).

The Landschaftspark, Duisberg-Nord (Figure 9.1.9) was developed from former industrial workings, and is considered a pioneering approach to urban regeneration and nature, now being a hotspot for

FIGURE 9.1.8 The New York High Line: 'Retrofitting the city with nature: the High Line New York: this project captured the public's imagination and helped redefine and globally influence what urban green space can be; it demonstrates how quality city space can positively utilise obsolete city infrastructure and also how a project of this scale can be successfully managed by the local community' (Cities Alive: Rethinking Green Infrastructure).

Source: Image credits: Lee Good

FIGURE 9.1.9 Landschaftspark, Duisberg-Nord (Peter Latz + Partner): 'The existing patterns and fragments formed by industrial use were taken, developed and reinterpreted by interlacing existing fragments into a new landscape' (Cities Alive: Rethinking Green Infrastructure).

Image credits: Raimond Spekking-CC BY-SA 3.0 (via Wikimedia Commons)

biodiversity in the region. The park demonstrates the healing hand of nature with diverse habitats, including rare species, and is divided into different areas, with planting being influenced by natural colonisation as well as the design of intimate gardens. The scheme recognises the value of the site's current condition, and uses phytoremediatiation to reduce the significance of pollution while embracing the memory implicit in industrial heritage (see https://en.wikipedia.org/wiki/Landschaftspark_Duisburg-Nord).

Conclusion

Cities Alive demonstrates how the urban environment is becoming increasingly understood as an ecosystem, delivering socio-ecological outputs across the entire range of situations. It recognises nature as the

FIGURE 9.1.10 Garden Bridge, London: 'Garden Bridge, London proposed by Heatherwick Studio, Arup and Dan Pearson studio, will be replete with grasses, trees and plants, offering new green space across the Thames in the very heart of London' (Cities Alive: Rethinking Green Infrastructure). The design offers a powerful ambition and vision regarding future city landscapes.

Source: Image credits: Arup

key driver of the system, presenting a range of precedents as examples of state of the art and best practice. Seen as at the forefront of urbanism, these address the diverse range of existing and new niches that are receiving attention to demonstrate current thinking and underpin future city development (Figure 9.1.10).

> Urban spaces are inextricably linked to the environment of our planet and there is a compelling case for building new cities and retrofitting or expanding existing ones in ways that work in harmony with nature by linking the ecological and human systems. The city can become an ecosystem that embeds nature and people as equal partners to help rebalance growing urbanism. The objective is to achieve a city that works in equilibrium and balance with nature, rather than against it to address the current excesses of urban living . . . At present, most cities woefully fail to achieve any sort of balance between people and nature.
>
> *(Cities Alive: Rethinking Green Infrastructure)*

9.2 Case Study 2 – Providing Local Green Infrastructure: the Case Study of Warndon Villages and the Warndon Project

This case study concerns the Warndon Project, initiated in the late 1980s, which involved an area of 800 acres of countryside situated on the outskirts of Worcester (UK), a location targeted for urban development. The project was highlighted at the time, since it concerned recognition of the landscape as requiring a high level of protection and assimilation into its new urban context in a way to ensure that its integrity remained and where it would serve the new community. The landscape had a distinctive character containing many important features of natural and cultural heritage, places of high biodiversity, hedgerows, ponds and ancient woodlands, as well as buildings and ancient trackways. Accordingly, development involved a strategy, agreed by all stakeholders, with a partnership between the City Council, the two major developers and the Countryside Commission, to safeguard and enhance the environment and thus make it available as a multi-purpose resource to the new residents.

The Warndon Local Plan established the concept of urban villages set among an extensive framework of landscaped corridors, and a landscape master-plan based upon ecological surveys. The key features of the project are reviewed here both in the context of the original project, but also regarding how the 'vision' is now being further explored and consolidated by Warndon Parish Council. The Warndon Project is particularly interesting not just because of its history, but because it represents an ecocentric approach to development, with an environment that demonstrates a high-quality biophilic profile that satisfies many of the requirements for everyday contact with nature. The information presented here offers an insight into the processes involved, and the landscape character that has matured over the years, to establish precedents that might be followed in other situations.

Essentially recognising the power of nature and viewed in the context of current discussion, it is clear that the project adopted an approach which was underpinned by a biophilic focus, the presence and accessibility of nature, establishing a balance between built and natural forms to create a powerful socio-ecological resource that satisfies many requirements, including multiple opportunities for immersion in nature, opportunities to get away from 'the routine', mental restoration, and encouraging physical activity. Specific support programmes are in place to encourage healthy activity based on walking, including an extensive network of pedestrian and cycling 'green-ways', as well as practical conservation activities and underpinning mechanisms to encourage environmental stewardship. Most recently, the vision for Warndon has been revisited and rejuvenated in the development of a 'neighbourhood plan' aimed at safeguarding and extending the green network, including the unique features of its biodiversity (Box 9.2.1). Once approved, it has statutory planning status (Andy Taylor pers com., 2022).

> Future developments within the parish of Warndon should maintain their balance between the residential, business, environmental and heritage areas of the parish, in keeping with the principles embodied in the Warndon Local Plan (1988) – the blueprint for the layout of the parish.
>
> *(Warndon Parish Neighbourhood Plan Consultation Summary)*

The Original Warndon Project

The project was a partnership between the local authority, the Countryside Commission and the developers, with the underpinning agreement to safeguard this heritage and make it available as a publicly accessible resource. In particular, housing was planned to fit within the existing landscape framework, the project being coordinated by a dedicated project officer. The involved change in land use, from farmed countryside to housing, thus moved features, including woodlands, meadows, ancient trackways and historical buildings, from a rural context to urban. Collaboration between stakeholders meant that all major green and heritage features were retained and extended to maintain and enrich local landscape character. This included new woodland planting and adding new green infrastructure to contain the development, forming corridors along transport and pedestrian routes which offer substantial scope for immersion in nature (Figures 9.2.1 and 9.2.2). Now mature, these plantings show an interesting system of green infrastructure together with the inclusion of wilder places which, overall, embrace free access, walking and enjoying the power of nearby nature as key attributes.

Warndon villages derives its name from the several original village locations and, as indicated, the area is served by a parish council; much information can be found on their website (warndonparishcouncil.org)

FIGURE 9.2.1A & B Pedestrian and cycle green-ways. Now mature, these are an impressive feature of the Parish, with a unique landscape structure capable of immersing the residents in green space.

and that of the Worcestershire Community Trust. For example, the Trust promotes physical activity through walking via Warndon Walkers, 'a walking group that meets at the Warndon Hub every Thursday. It is a free activity where you walk at a relaxed pace for one hour' (www.worcestercommunitytrust.org.uk/project/warndon-walkers). Related information includes details of walking routes, 'Warndon Villages Woodland Nature Trails' and 'Heritage Walks', while additional published information developed by a local resident Barbara Hopper includes 'Walks in the Parish' and 'Walks from Lockdown' (see Parish Council Newsletter Winter 2020, warndonparishcouncil.org). The current Green Biodiversity Strategy is 'to allow nature to find its own way' (see Warndon Parish Council Newsletter Spring 2020, warndonparishcouncil.org). The area also showed innovation regarding nature conservation in installing special bat-friendly lighting where the A4440 impinges on Warndon Wood Nature Reserve (www.worcestershire.gov.uk/news/article/1890/the_first_bat).

> Warndon Villages was a vision of a planned townscape sympathetic to the historical places which it would encompass. It was the largest new urban development in Great Britain in the 1980s and its special wooded character is greatly valued by residents and visitors alike.
>
> *(Warndon Parish Council, Warndon Villages Heritage Trail; see warndonparishcouncil.org, accessed 12-03-2022)*

Significantly, the development was set in an agreed landscape strategy and supporting initiatives. For example, publicity leaflets about 'The Warndon Project' given out to new houseowners introduced them to the natural attributes of the area, while the same information encouraged exploration by walking and gave advice to new residents on how to improve the wildlife value of their own gardens. These actions have in large continued, now being promoted by the Parish Council, but also by individual residents and voluntary groups. The new planting has matured and can be explored in relation to the population that it serves and how this same population is safeguarding it. The area demonstrates interesting landscape typologies and related actions which serve to define the quality of the environment and associated measures. Essentially, these actions can be regarded as identifying, safeguarding and increasing the resilience of the biophilic features (Figures 9.2.2 and 9.2.3).

FIGURE 9.2.2 Trotshill Field: one of the original natural features retained as a wild place.

Source: Image credits: Andy Taylor/Dawn Merriman

Warndon Villages Highlights and Precedent Features

- An area of 800 acres developed from countryside in the late 1980s to meet the expansion of Worcester into the 21st century.
- Collaboration between Worcester City Council, the Countryside Commission* and two consortia of developers, coordinated by a dedicated project officer.
- 4,500 new homes as well as a large industrial park, shops and community facilities, including a new road network.
- Landscape master-plan based on natural features, contours and a landscape framework.
- Retention and incorporation of natural heritage and historical features within the new planned suburban structures.
- Ancient routes perpetuated as cycleways linking to walking routes and bridleways.
- Walking and free access as underpinning criteria with the provision of 'green-ways'.
- New woodland plantings to extend existing woodlands.
- New green infrastructure using a mix of species to complement existing character.
- Planting palette which reflects the native woodland, but enriched with some key evergreen trees and shrubs to ensure resilience and accommodate/balance built forms.

FIGURE 9.2.3 Trotshill Wood: situated in the heart of the new residential development, highly accessible woodland, very popular. There are now signs that the native flora is spreading to adjacent new plantations.

- 'Welcome to Your new Home' leaflet including information on woodlands, wildlife, trails and guidance on garden planting for wildlife.
- Strong green corridors established along pedestrian routes involving trees, shrubs and new hedgerows to enable immersion in 'green'.

- High-quality nearby nature, including nature reserves, which facilitate deeper immersion and restorative experiences.
- Recognising and retaining important views into the wider natural landscape.
- First example in the UK of special 'bat-friendly' road lighting.
- Residents' interests now represented by the Warndon Parish Council.
- Parish Council Newsletter includes features on the environment, heritage and 'About and Around', updates on recent actions.
- Related parish council activities which encourage stewardship of the landscape.
- A volunteer environmental group active in the area (and throughout Worcester) which undertakes conservation and other work in partnership with the city council.
- Events and competitions, including allotments competitions.
- The recent development of a Neighbourhood Plan consolidating the importance of the landscape, biodiversity and key ecological sites within the context of the original landscape vision, stating management aims and related principles for enriching and protecting the landscape character, biodiversity and providing planning guidance.

(Information summarised from Warndon Parish Council website, including newsletters and archived material regarding the Warndon Project and the Warndon Parish Neighbourhood Plan Consultation Summary Booklet.

**Note that The Countryside Commission was abolished in 1999 and eventually incorporated into the statutory body Natural England.)*

BOX 9.2.1 ECOCENTRIC AND BIOPHILIC FEATURES INDICATED IN THE PROPOSED WARNDON PARISH NEIGHBOURHOOD PLAN

- **The draft Plan**, currently at consultation stage, developed by the parish council, seeks to turn community aspirations into planning policies to inform future developments, setting out a range of objectives which emphasise, the environmental characteristics of the area, their importance, protection and enhancement, including:
- **Landscape character**: measures to maintain and reinforce Warndon Parish's natural environment and landscape character

will be promoted wherever possible; measures include having regard to conserving and enhancing the landscape character, its beauty, biodiversity and amenity value, the setting character, appearance and cultural heritage. In keeping with existing settlement patterns, maintain the concept of villages and hamlets set within a framework of green corridors; protect key landscape features and biodiversity, including hedgerows, trees, woodlands, orchards and ponds, adding to the natural assets where opportunities are available; safeguard open countryside and cherished public views.

- **Biodiversity**: development proposals that result in a net loss of biodiversity will not be permitted; new proposals must protect and enhance biodiversity by protecting nationally and locally designated sites, protected species and ancient or species rich hedgerows, grasslands, woodlands, trees and ponds. Preserving local ecological green networks; extending existing and creating new ecological green networks where opportunities exist; promoting the preservation, restoration and enhancement of wildlife habitats, the protection and recovery of priority species; provide mitigation measures and the re-provision of habitats for species lost through development; provide a measurable net gain in flora and fauna.
- **Trees and hedgerows**: recognises that trees and hedgerows are an integral part of the Neighbourhood Plan area, ancient trees and hedgerows of arboricultural and amenity value should be retained, development that damages or results in their loss will not normally be permitted; proposals to be accompanied by tree surveys, trees to be protected during development; removed trees to be replaced at a ratio of at least 2:1, preference for locally occurring species, standards set for tree planting requirements for new residential and other situations; requirements to secure planting in other situations to mitigate any tree or hedgerow loss, including meeting maintenance obligations.
- **Green routes for travel, footpaths and cycle paths**: support development that would result in improvements to footpaths and cycle paths, but not to conflict with other policy aims; support to proposals that improve and extend the existing green routes, footpaths and cycleways and that encourage active travel; new development must demonstrate how walking and cycling opportunities have been prioritised and connections made to existing routes;

the use of Neighbourhood Community Infrastructure Levy Funds and section 106 planning agreements will be used to enhance and expand these routes; access to footpaths and cycle-paths should be unobstructed with appropriate signage; native planting should be retained and, where possible, enhanced to retain wildlife corridors, support local biodiversity and maintain the traditional landscape character; proposals which adversely affect existing walking and cycling routes, or fail to encourage appropriate new walking and cycling opportunities, will not be supported.

Notes

1 The entries are compiled from information in the Warndon Parish Neighbourhood Plan, Consultation Summary Booklet, but for precise wording and full information, reference should be made to the original document: see https://consultations.rcaregeneration.co.uk/warndon-neighbourhood-plan-21
2 The Plan also includes additional policies regarding employment in the parish, retail and community facilities.
3 The entries here specifically represent a summary of selected information taken from the plan and are intended to demonstrate the ecocentric approach and biophilic features of the policies; they use the phraseology adopted in the plan but are presented in a more concise format.
4 The community Infrastructure Levy is a sum of money charged by the City Council to a developer to be used towards schemes which will benefit the wider community. Section 106 agreement is a legal agreement between the council and developers to deliver a range of benefits, typically of wider impact than within the specific confines of the development site.

Conclusions

In re-visiting Warndon Villages, it is possible to see how the character of the area has developed since housing development first began, where the focus has been on the quality of the environment, and that the location has an impressive and resilient system of green infrastructure. In re-appraising its landscape, the parish council has reaffirmed the key principles and visions in the original plan, and supports the local community in disseminating information and facilitating its use, now further establishing its importance and protection within planning legislation in the form of a Neighbourhood Plan.

Actions are seen to demonstrate a co-creation approach, involving working with the community in safeguarding the environment, but also in supporting people in gaining maximum benefits from its use. The landscape now forms a network of connected natural and biophilic spaces which show the types of features being advocated for the effective development of green infrastructure which can be experienced on a daily basis, but which also offers opportunities for deeper engagement. These naturally support walking and cycling activities, i.e. involving moderate exercise, but also accord with the ways that the landscape is perceived, that it needs to be resilient and provide diversity regarding interaction. Overall, it established important precedents which can be applied across urban situations concerning biophilic-focused planning of residential areas involving green infrastructure, and demonstrates that such initiatives should, ideally, be underpinned by appropriate local environmental management plans supported by relevant planning legislation.

9.3 Case Study 3 – Biophilic Streets – Initiatives on Shifting Street Paradigms

This case study takes up the theme of biophilic urbanism to explore the idea of biophilic streets. It presents several examples of streets which have been re-imagined as multifunctional biophilic places set aside as special locations, where natural sounds and natural qualities are prioritised, and which provide rich opportunities for immersion in nature, fostering connections among people and communities. Conceived to 'harbor native plants and biodiversity, that collect and treat stormwater, and where pedestrians can experience intimate contact with nature as part of their daily routine' (Beatley, 2011, p. 99), biophilic streets would present nature in ways that it can be appreciated and experienced at different levels of personal and social interaction.

Biophilic streets thus embrace ideas around the importance of walkable, dynamic and habitable neighbourhoods, and assimilates the thinking of urbanists such as Jane Jacobs and Jan Gehl, but also challenge mainstream thinking, moving the street from what many have become, essentially that of a purely functional transport corridor, to one that is capable of being transformative in terms of urban street life. Fundamental are the ways that they provide high-quality experiences as intergenerational and multifunctional public spaces that families and communities can relish together: places where neighbourly interactions occur, spaces for playing, socialising and getting around more independently, where wandering and exploration of the environment is possible and pleasurable without the need to travel long distances (Arup, 2017).

The idea of biophilic streets has certain parallels with the idea of play streets developed at the beginning of the 20th century. Due the lack of specified play areas, large cities such as London and New York were campaigning for children to play in the streets and for these to serve as playgrounds. Vehicular traffic and parking were limited, and cities became familiar with signage such as 'Play Street, no thru traffic, no parking in this block', 'play street vehicles banned' or 'play street no vehicles from sunrise to sunset'. The visions behind biophilic streets thus re-conceptualise the streets as active biophilic places where nature comes to the fore, and as such, promotes our understanding of what urban environments could be, including how urban space is conceived, promoted and utilised in the most efficient and environmentally friendly way. Biophilic streets apply understanding of how we need to and can associate with nature in our everyday activities, and vision the future of the street, where it is easy to see streets repurposed as an interlinking biophilic network delivering benefits in multiple ways.

Champs-Élysées, Paris

The collaborative study for *Re-Enchanting the Champs-Élysées* proposes a transformation where more life is brought into the avenue and its neighbourhood areas (Figures 9.3.1 & 9.3.2). It is prompted by the realisation that an important artery of the city was losing its identity, no longer being suited to the needs of Parisians, and needed to be re-invented and re-appropriated. The vision for 2030 is underpinned by four principles (PCA-STREAM, 2020):

1 Decrease the mobility nuisance: this will include an increase in pedestrian-friendly areas, reducing vehicle traffic by 50%, and restoring connections between the Tuileries Garden and the Champs-Élysées Gardens and other important public spaces; additionally, it will consider a range of strategies for mobility such as biking lanes, shared biking and transport networks, as well as installing low-noise road surfacing, change to an electric vehicle fleet and the significant increase in the urban forest will result in a dramatic reduction in noise pollution, fine-particle air pollution and CO2 emissions by 2030.
2 Invent new urban uses: this will involve reinventing the Champs-Élysées Gardens through the use of temporary and reversible

FIGURE 9.3.1 Study for the redevelopment of the Champs-Élysées: Haut des Champs Élysées salon végétal.

Source: Image credits: PCA-STREAM.

FIGURE 9.3.2 Study for the redevelopment of the Champs-Élysées: Haut des Champs Élysées salon végétal variante.

Source: Image credits: PCA-STREAM.

installations, playground structures and spaces for outdoor sports and wellness facilities, in addition to sustainable and inclusive food kiosks and pop up-stores, a coordinated programming of cultural and scientific events throughout the year, multifunctional spaces for concerts, open-door cinemas and art exhibitions.

3 Rethink nature as an ecosystem: the strategy includes creating planted 'lounges' along and adjacently to the avenue, increasing the urban tree canopy, understorey and soil permeability; it takes into consideration water sensitive design with rainwater harvesting, bioswales and buffer strips to increase water infiltration, and retention basins to filter pollutants. The aim is to have a living ecosystem that generates shaded spaces, attracts wildlife and thus increases biodiversity, as well as resulting in other environmental benefits associated with air quality, cooling and water infiltration.
4 Data as a tool: considering digital infrastructure to help regulating the urban metabolism, including monitoring flows ranging from transportation and food to storm water and information, with wider economic and environmental benefits for the area.

Alfred Place, London

Alfred Place (Figures 9.3.3 & 9.3.4), an underused road in London, has been transformed into a park, helping to tackle issues associated with the lack of shared green spaces, poor biodiversity, poor air quality and flooding. It offers a meandering path edged with seating that people

FIGURE 9.3.3 Alfred Place, a general view of meandering paths, sitting areas and new planting beds.

Source: Image credits: studio.maple/LDA Design.

FIGURE 9.3.4 Alfred Place, lounging lawns.

Source: Image credits: studio.maple/LDA Design.

use at lunchtime, and small lawns and planting that support biodiversity, user play and interaction. The project is part of the London West End Project, which is investing in creating new and improving existing parks, pockets parks and streets, to support the health and well-being of communities living in the Tottenham Court Road area, with places to walk, cycle, meet, relax and play (LDA Design, 2022; see also Camden's West End Project).

London also embraces a 'Healthy Streets Approach', a concept being advocated by Transport for London, which aims to help people to adopt healthier behaviours towards active travel, encouraging walking, cycling and use of public transport. The approach is intended to promote attractive places to live, work, play and do business, with 80% of all trips within London to be made by walking, cycling and public transport by 2041 (Transport for London, 2017).

The High Line, New York

The High Line in New York (Figure 9.3.5) is known internationally as a ground-breaking example of repurposing redundant post-industrial urban infrastructure. Now forming a contemporary linear landscape, it provides an urban microclimate of textured and colourful habitats that engage its users in strong individual and social sensory experiences throughout the year, in which people can explore their individual choreographies of walking, slowing down and stopping to appreciate the views and engage with the city, through its various outdoor rooms that allow for contemplation and deep reflection.

São Paulo, Brazil

São Paulo (Figure 9.3.6) reimagined its streets by converting part of street spaces into 'parklets', consisting of parking space that has been transformed into a greener area for people to sit, meet up, linger and participate in the social life of their neighbourhood. The ideas were one of the responses to the highly congested streets of São Paulo in an effort to make the city more human. This type of tactical urbanism would later inspire many other cities as a response to the challenges of COVID-19, which involved changing work patterns and the need for more local attention to the environment; while parklets are considered a temporary response, they do suggest attractive options for more permanent consideration.

FIGURE 9.3.5 The High Line engages people with nature and the city.
Source: Image credits: Lee Good.

Avenue Mont-Royal, Montreal

The Avenue Mont-Royal (Figure 9.3.7–10) is also a temporary intervention in a street that is usually used for vehicular transit in Montreal, during the summer months being transformed into a 2.5km linear park

FIGURE 9.3.6 Parklet as public spaces. Features include planting beds, sitting and bike parking areas.

Source: Image credits: Wagner Tamanaha from Sao Paulo, Brazil, CC BY-SA 2.0 <https://creativecommons.org/licenses/by-sa/2.0>, via Wikimedia Commons.

FIGURE 9.3.7 The forest bench.

Source: Image credits: Latrompette Studio.

FIGURE 9.3.8 Food garden.

Source: Image credits: Latrompette Studio.

FIGURE 9.3.9 Edible gardens with inner trail.

Source: Image credits: Latrompette Studio.

that interconnects the avenue to its five parks, from Mont-Royal to Baldwin Park. The pedestrian route is dotted with artistic interventions and installations focusing on relaxation, playfulness, vegetation and staying cool, all aimed at encouraging an immersive experience, walking and meeting. In 2022, more than 1000 seats, forest benches and streams where children can play and race small boats, were installed alongside associated planting.

FIGURE 9.3.10 The water stream provides cooling and opportunities for children to play.

Source: Image credits: Latrompette Studio.

As a biophilic place, the Avenue demonstrates substantial innovation and unique responses to the extreme summer temperatures, including refreshing stops, the breeze pool, the mist oasis and the water jets of the 'Place Fleurs-de-Macadam'. Restaurants and cafes, plus edible gardens and associated workshops on urban agriculture, outdoor cinema, exhibitions and other events, add to the multifunctional character and to the sense of place of the avenue. Importantly, the lifecycle of the materiality involved in this transitional landscape was thoroughly considered by reusing, transforming, repurposing, and recycling materials from the previous years and through replanting the vegetation throughout the neighbourhood (Odace Événements, 2022).

Superblocks, Barcelona

The 'Superblocks' concept is a strategy and a model of city transformation aiming to give priority to pedestrians and cyclists by reclaiming streets and other public spaces from cars (Figures 9.3.11 & 9.3.12). It highlights a paradigm shift for the city of Barcelona, involving a partnership between Ajuntament de Barcelona, The Urban Ecology Agency of Barcelona (BCNecologia) and residents in a vision that considers the role of streets in making local communities healthier and adopting more active lifestyles. The Superblock is a people-centred strategy which plans to include a total of 503 Superblocks across Barcelona with a range of macro to micro-scale interventions (Ajuntament de Barcelona, accessed July 2022). It is estimated

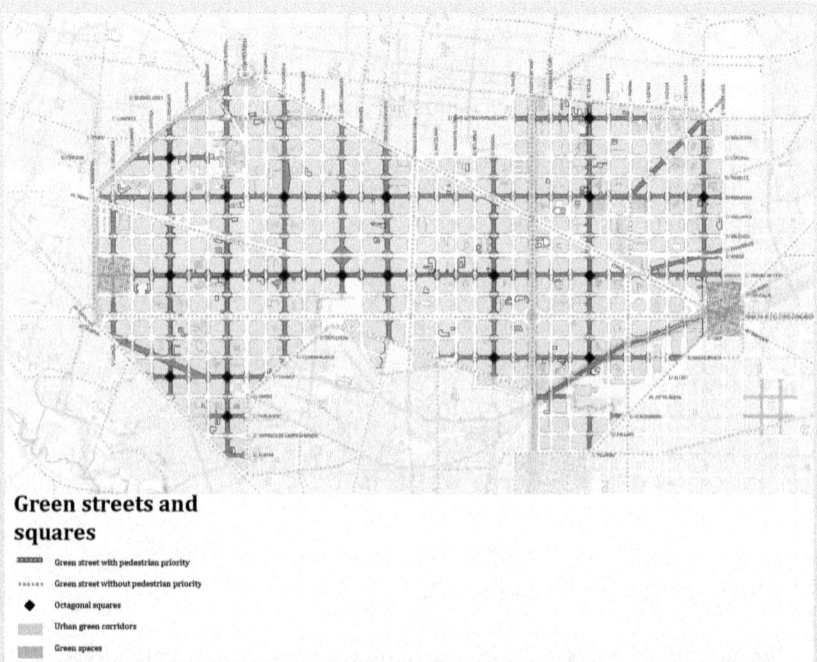

FIGURE 9.3.11 Proposal of a network of streets and squares for the Eixample district: 33km of green streets, 21 squares, increase of 3.9 ha of pedestrian space, 6.6 ha of urban green, proximity of 200m for local population.

Source: Image credits: Ajuntament de Barcelona (Accessed July 2022).

FIGURE 9.3.12 Aerial view of the proposal for a new central square with trees and spaces for people to stroll and stay at the junction between Consell de Cent and Comte Borrell.

Source: Image credits: Ajuntament de Barcelona (Accessed July 2022).

that 60–70% of the road space will be released from car traffic (Staricco & Vitale Brovarone, 2022). As a whole, the strategy provides a major contribution to the creation of biophilic streets that allow citizens to be immersed in an integrated and continuous matrix of urban nature, stitching the built and the natural environment together; it is thought to prevent at least 667 premature deaths each year (Mueller et al., 2020).

Main characteristics:

- Reorganises neighbourhoods by repurposing the street network and intersections of the blocks in order to redesign how the public space is used and activated.
- Involves creating healthier, fairer and safer public spaces which are transformative to the lives of local communities and economies.
- Limits the impact of the road network, reduces dependency on cars, and improves neighbourhood liveability by providing equal opportunities to access a higher percentage of green community spaces, walkable and cycling spaces, as well as linking to a highly connected public transportation system.
- Fosters urban biodiversity with microhabitats which attracts avifauna.
- Provides biophilic encounters through added purposeful biophilic components that allow new visual connections with the natural world and natural processes whilst walking, promenading, lingering, playing and meeting with others.
- Provides a playful intergenerational space, increasing social interaction and social cohesion.
- Supports infants, toddlers and their caregivers by being implemented in dense and mixed-used neighbourhoods with higher number of nurseries, schools and community centres, offering playscapes, shade and urban furniture that supports meeting and remaining.
- The new street network provides larger and safer public spaces, offering opportunities for outdoors activities in better air quality conditions.
- Makes use of urban furniture, graphic applications and paving patterns to create an identity and sense of place.
- Creates new social and urban identity.

See more at:

Ajuntament de Barcelona: https://ajuntament.barcelona.cat/superilles/en
The Urban Ecology Agency of Barcelona: www.bcnregional.com/

Pathways to Successful Biophilic Streets

It is clear that such projects demonstrate a paradigm shift towards creating streets that are biophilic and people orientated, being underpinned by visions that support pathways to the health and well-being of communities, but also help in future-proofing against environmental and related change. The development of biophilic street systems provides direct, positive and everyday contact with nature in embracing interventions that express new urban systems which activate the street as a public and democratised space. These bring renewed social and cultural life to streets, while fostering neighbourhood identity and increasing social capital. As such, they demonstrate a series of precedents applicable to many situations across the urban scene, which highlight that, as easily accessible spaces, biophilic streets provide urbanites with everyday access to green spaces, being particularly significant in highly dense neighbourhoods and to those without access to other spaces (Kellert & Calabrese, 2015).

Precedents and principles:

- Co-creation of public space: must involve both the communities and local authorities/agencies in the creation and implementation process underpinned by trust and democratic actions. User input will contribute to environmental stewardship and resilience of those interventions.
- Scale and scaling up: involves intervention at different levels, whilst reinforcing the pedestrian scale for inviting community interactions. Small interventions can be scaled up to create continuous and interconnected biophilic streets, squares and gardens with high overall impact.
- Multifunctional: offer a high diversity of functions for everyday activity, including various forms of sustainable mobility, commerce, access to nature and playscapes.
- Mobility: need to limit vehicular traffic and encourage journeys on foot and by bicycle, which help to promote healthier and active lifestyles.
- Biophilic elements: abundant in nature and biodiversity (e.g. trees, planted borders, raised beds, edible trails, living walls, rain gardens, waterscapes) bringing new perspectives on how natural and ecological systems can be accommodated in cities, and their benefits, including the communities' experiences of nature, exposure to the natural world and related sensory interaction, listening in, seeing, smelling, tasting and haptic interactions.
- Educative: should offer opportunities for teaching, learning and play.

- Adaptable furniture and structures: which will allow for flexibility to suit community and user needs, individuals and groups, respect different paces and modes of interaction, slowing down, lingering and picnicking.
- Food: food in the forms of temporary or permanent eating places and urban growing spaces attract street life and people of all ages.
- Water sensitive: manage water so it creates opportunities for playing (e.g. water rills and narrow streams), cooling, irrigation, filtering and tackling surface flooding.
- Street art: use street art, graphic applications and paving patterns to bring a stronger neighbourhood identity.
- Tackle environmental issues: contribute to reducing noise, air pollution and heat island effect, and respond to climate and biodiversity emergencies.

Biophilic streets thus feature readily accessible and abundant nature experiences that provide sensory, physical and mental stimulation and contribute to the neighbourhood sense of identity, delivering opportunities for walking, sitting, meeting and talking to others, playing, relaxing, watching and working, thus leading to more meaningful encounters with the urban context. They demonstrate particular value and innovation in humanising streets and promoting health and well-being benefits to neighbourhoods and across the city.

References

Ajuntament de Barcelona, Superillas. https://ajuntament.barcelona.cat/superilles/en/

Arup. (2017). *Cities Alive. Designing for Urban Childhoods*. London: Arup.

Beatley, T. (2011). *Biophilic Cities. Integrating Nature into Urban Design and Planning*. Washington, DC: Island Press.

Kellert, S., Calabrese, E. (2015). *The Practice of Biophilic Design*. www.biophilic-design.com.

LDA Design. (2022). *Alfred Place Gardens – From Tarmac to Treasure*. www.lda-design.co.uk/kindling/news/alfred-place-gardens-from-tarmac-to-treasure/

Mueller, N., Rojas-Rueda, D., Khreis, H., Cirach, M., Andrés, D., Ballester, J., Bartoll, X., Daher, C., Deluca, A., Echave, C., Milà, C., Márquez, S., Palou, J., Pérez, K., Tonne, C., Stevenson, M., Rueda, S., Nieuwenhuijsen, M. (2020). Changing the urban design of cities for health: The superblock model, *Environment International* 134, 105132. https://doi.org/10.1016/J.ENVINT.2019.105132

Odace Événements. (2022). *Mont-royal Avenue Turns into a Linear Park.* Montréal: Odace Événements.
PCA-STREAM. (2020). *Exhibition Champs-Élysées. History & Perspectives.* Paris: PCA-STREAM.
Staricco, L., Vitale Brovarone, E. (2022). Livable neighborhoods for sustainable cities: Insights from Barcelona, *Transportation Research Procedia* 60, 354–361. https://doi.org/10.1016/J.TRPRO.2021.12.046
Transport for London. (2017). *Healthy streets for London. Prioritising Walking, Cycling and Public Transport to Create a Healthy City.* https://tfl.gov.uk/cdn/static/cms/documents/healthy-streets-for-london.pdf

9.4 Case Study 4 – Delivering Environmental Justice: The Birmingham City of Nature Plan

This case study considers Environmental Justice, specifically examining issues concerning the rights and opportunities to access nature places, looking at the example of how Birmingham (UK), through its City of Nature Plan, is developing an agenda to respond to environmental injustices across the city. The issue is relatively simple: access to green space and related environments concerns opportunities to experience the benefits of being in nature, encouraging walking and other forms of physical activity and should be underpinned by the appropriate provision of the same across the city. However, the reality in many cities is that green space provision is not equally distributed, of appropriate quality or available to all, and requires initiatives to address the situation.

There is strong evidence that less green neighbourhoods have more health inequalities associated with deprivation (Mitchell & Popham, 2008), with negative effects on the quality of life and on life expectancy. For example, studies in the US found that communities and families of colour and/or low income have significantly less access to nature (Nature Quant, 2021); in effect, some neighbourhoods are severely nature-deprived. In financial terms, outdoor physical activity in nature can deliver health benefits, in the UK worth £2.18 billion each year (White et al., 2016, see also Natural England, 2009). However, the effects of nature deprivation have broader and long-term consequences, particularly on children, regarding disconnect from nature and nature play and obesity-associated issues.

Birmingham has not been immune to these challenges in terms of the health and well-being of its inhabitants. Health inequalities show a direct link to poverty in the city, with 408,000 Birmingham citizens living in the top 10% most deprived areas in England. Birmingham wards have one of the highest discrepancies in terms of life expectancy in the UK, and high levels of inactivity (more than 80% of the population) with around ¼ of Birmingham's children being overweight/obese when they start school, rising to 40% by the time they leave primary school (data retrieved from The Active Wellbeing Society, 2019). It is these wards that also have the poorest distribution of green space and green infrastructure. For example, a survey shows that, in Birmingham, the most affluent wards have substantially more tree cover than the more deprived areas (Chandler, 2021) and that, in the more deprived wards, the population at large faces substantial barriers in accessing green spaces or making full use of facilities on offer, a situation which can be particularly marked for certain groups (Birmingham City Council, 2022).

The notion of Environmental Justice is broad and well defined in the literature, and includes the rights to experience healthy environmental conditions and to freely benefit from access to green and related places irrespective of circumstances (see, for example, IGBC, 2022). Cities contain a variety of green infrastructure typologies, but typically it is the public open spaces which are of particular importance, and in many situations are the only places available offering nature or biophilically related recreational facilities; it is these rights and associated issues that are addressed in the plan.

The Birmingham City of Nature Plan (Birmingham City Council, 2022)

The Birmingham City of Nature Plan is an integrated strategy developed to challenge current management models of green environments, which often rely on silos within a single profession/department, and to move on to a cross-system thinking from which nature and environmental justice are fundamental parts in resolving many of its urban challenges.

The City of Nature Plan was developed through the Birmingham Future Parks Accelerator Project, which is a UK programme that aims to shape a better future for public spaces such as gardens and urban parks. It involves initiatives across nine cities, including Birmingham. The programme, which is a collaboration between the National Lottery Heritage Fund, the National Trust and the Ministry of Housing and Local Government, recognises that whilst urban parks have been neglected over the years, they can provide invaluable benefits for communities, nature and sustainable development. The programme aims to help local authorities to 'put green spaces at the heart of their decision making and support them in engaging with communities as well as developing financial strategies and capacity to manage such spaces' (www.futureparks.org.uk/about).

The Birmingham City of Nature Plan (Birmingham City Council, 2022) is a 25-year vision that considers the sustainable provision, management and use of green and blue spaces in the city, connecting people to nature, and tackling the challenges of inequity, health and well-being, nature recovery and climate change. It includes the on-going development of an Environmental Justice map for Birmingham*, which not only considers the parameters of distance and access to green spaces, but also addresses flood risk, urban heat island effect, the UK Index of Multiple Deprivation and the Excess Year Life Lost. The plan is underpinned by cross-system thinking and the vision to bring nature to the fore in resolving many of the city's challenges by addressing environmental injustices and helping to tackle deprivation issues. The vision is supported by five citywide themes that put nature at the core, and put emphasis on wider decision-making in supporting a circular economy approach:

1 A green city: ensuring the city enables nature to fully integrate city's life by having green and blue infrastructure that is safe, clean and sustainably managed; it includes increasing green spaces and tree canopy cover across the city, and placing an emphasis on urban development schemes that have a strong green focus.
2 A healthy city: ensuring every citizen in Birmingham can access green spaces to improve their health and well-being as part of the foundations of a good life, offering residents safe and attractive cycling and walking paths.
3 A fair city: ensuring that the city offers its highly diverse and distinctive communities the same opportunities in accessing good-quality green space independently of where they live, with the pledge to create 400 green spaces prioritizing those areas and communities who need them the most; it also proposes to ensure fair access to green jobs and that [their] workforce reflects [the] diverse communities.
4 An involved city: centring stewardship on people and communities to love and protect their green environments, aiming that every child in the city will grow up experiencing nature.
5 A valued city: ensuring that the city appreciates and captures the value of nature as a drive for growth and as an important component of business investment in the city, promoting a range of mechanisms to support and sponsor communities.

FIGURE 9.4.1 Canals in Birmingham City Centre are a stage for biophilic encounters and multifunctional activities, kayaking, walking, sitting, boating, eating and drinking.

FIGURE 9.4.2 The Canals in Birmingham are an extensive blue infrastructure with potential to connect communities together and help respond to environmental injustices.

Action Plan and Levelling up

The action plan to 'level-up' environmental justice in the city includes setting a new benchmark of quality for all Birmingham's green spaces, by assessing these spaces at local level against the Birmingham Fair Parks Standard, an environmental justice tool which examines if the green space is:

- Fair (is it welcoming, accessible, clean and safe?)
- Green (are there different trees and plants, are there habitats for wildlife, is it managed sustainably?)
- Healthy (are there walking routes, quiet areas, activities, is the park used for social prescriptions; does it have play value?)
- Involved (can you find out what's happening in your park, can you influence what is happening?)
- Valued (do we know the worth of what the park provides, is that shared, are there ways to raise extra funds?)

The approach will be first piloted in five open spaces in Bordesley and Highgate Ward before it is taken to other wards which are deemed to suffer most from nature deprivation and environmental injustices. By 2047, Birmingham aims that all its parks and green spaces will meet the Birmingham Fair Parks Standard and be able to be accessed, used and managed to their full potential, resolving the unequal access and distribution of green space. The Birmingham City of Nature Plan demonstrates the city's ambition and the intent to be green and sustainable, improving the connections between its residents, health and nature, and considering a response to climate change and biodiversity loss as issues central to its agenda. Significantly, in line with its new attitudes towards nature and the development of the City, in 2013, Birmingham declared its intention to become the first Biophilic City in the UK.

Birmingham's vision and plan recognises that local parks and similar spaces are the most important green assets for the communities which can make a substantial difference to health, well-being and community identity, and seeks to develop partnerships to ensure that places are well managed, accessible, fostering measures that support positive use and a sense of stewardship. For example, in 2019, in the Lozells ward (identified for being one of the most deprived areas and having the lowest male life expectancy in Birmingham), a Friends of George's Park community group was founded to 'activate, stimulate and look after this place'. The group engages all ages through a number of volunteer-led actions, from community litter picking to outdoor fitness classes,

FIGURE 9.4.3 Garrison Lane Park, Birmingham. Located in the highly deprived ward of Bordesley Green, and home to mature London planes and large grassed areas, this is one of the Birmingham parks planned for improvement under the Environmental Justice framework in order to suit the contemporary needs of its community.

FIGURE 9.4.4 Cannon Hill Park. Located in an affluent area, it attracts people from all over Birmingham all year round. It is thus considered an example of success for its inclusive nature, embracing a range of users needs that link well with the highly multicultural population of Birmingham, and offers a diverse programme of activities and spaces to engage with nature.

walking, activities for children and teenagers, planting flower beds, outdoor cinema and multicultural picnics. The impact for the community has proven to be vital for quality of life, in terms of physical and mental well-being, but also how it now promotes positive social and community interaction, which was of particular importance during the COVID-19 lockdown (Afzal Goni podcast interview Nov, 2021).

Conclusions and Highlights

Birmingham has faced up to the challenges regarding its communities and has recognised the role of nature as a supporting influence in terms of health, well-being and community. Set in the context of improving quality of life for all its residents, Birmingham has developed its City of Nature Plan as a long-term vision to develop the biophilic characteristics of the city, address nature deprivation and remove environmental injustices. In doing so, it demonstrates a vision which is underpinned by the power of nature and involves actions which set standards and involves local, on-the-ground initiatives to support parks and other places as vibrant nature-focused areas for the community, to tackle issues of nature deprivation and set a charter for environmental justice.

In becoming the UK's first Biophilic City, the Plan provides pathways and set standards which are likely to be followed by other initiatives. This involves partnerships and solutions that address the issue of nature deprivation across the city, and seeks to reach a situation where opportunities for contact with nature underpin much of its activities and identity. The plan involves envisioning a city future where we understand and work with the natural world, embracing a number of fundamental challenges and actions:

- **Challenges status quo**: by defining thought processes and new models that work outside of silos and puts nature at the core of the decision making.
- **25-year inclusive vision**: offers a clear timescale to become a distinctive 'City of Nature'.
- **Environmental Justice**: Birmingham is the first UK local authority to develop a tool to assess Environmental Justice, recognising diversity and considering access to good-quality green space and fair treatment of people of ethnic diverse communities.
- **People-focused**: education, advocacy and empowerment of communities by offering them the tools to access, experience and protect nature.

- **Connecting and restoring**: connecting communities and businesses through restoring nature spaces in the city, connectivity and access.
- **An Environmental Justice tool**: new benchmarks in Birmingham Fair Parks Standards.
- **Community, council partnerships:** specific initiatives to connect parks and people, support confident use.
- **Enhanced green infrastructure**: increase the range of safe, clean and sustainably managed green spaces offered (e.g. growing spaces and communal gardens) and the tree canopy coverage across the city to 25%.
- **Active travel**: integrated to align with green corridors and help improve air quality.
- **The Nature Recovery Network:** bringing stakeholders together in a partnership to encourage and manage wildlife-rich places, to increase and restore nature and the ecological value of habitats, in order to respond to three major challenges: biodiversity loss, climate change and well-being.

Note: * Birmingham Environmental Justice map available at ttps://www.birmingham.gov.uk/download/downloads/id/23230/birmingham_environmental_map.pdf

References

Afzal Goni (2021). *Episode One – WMNP award winner: Friends of Georges Park*. Birmingham: Birmingham City University. https://soundcloud.com/birmingham-city-university/episode-one-wmnp-award-winner-friends-of-georges-park?in=birmingham-city-university/sets/west-midlands-national-park-awards

Birmingham City Council. (2022). *Birmingham City of Nature plan*. https://naturallybirmingham.files.wordpress.com/2022/01/appendix-a-city-of-nature-document_cabinet-final.pdf

Chandler, N. (2021). *Deprivation and Urban Forest Inequality: Spatial Methodology for Identifying Spaces for Miyawaki Forests* (LAN7342: MA Landscape Architecture). Birmingham: Birmingham City University. Unpublished student paper.

Future Parks Accelerator. www.futureparks.org.uk/about

IGBC. (2022). *International Green Building Council: Urban Green Space, Public Health, and Environmental Justice*. https://internationalgbc.org/2022/01/22/urban-green-space-public-health-and-environmental-justice-the-challenge-of-making-cities-just-green-enough/.

Mitchell, R., Popham, F. (2008). Effect of exposure to natural environment on health inequalities: An observational population study, *Lancet* 372(9650), 1655–1660. https://doi.org/10.1016/S0140-6736(08)61689-X

Natural England. (2009). *An Estimate of The Value and Cost Effectiveness of the Expanded Walking the Way to Health Initiative Scheme 2009* (TIN055). http://publications.naturalengland.org.uk/publication/35009

Nature Quant. (2021). *Nature Access and Environmental Justice: Addressing Nature Disparities in the United States* (NatureQuant.com). www.nature-quant.com/NatureQuant-Nature%20Access_and_Environmental_Justice.pdf

The Active Wellbeing Society. (2019). *The Role of Design for Active Communities* (Talk delivered by Karen Creavin). Birmingham: Birmingham City University. https://theaws.co.uk/

White, M. P., Elliott, L. R., Taylor, T., Wheeler, B. W., Spencer, A., Bone, A., . . . Fleming, L. E. (2016). Recreational physical activity in natural environments and implications for health: A population based cross-sectional study in England, *Preventive Medicine* 91, 383–388. https://doi.org/10.1016/J.YPMED.2016.08.023

9.5 Case Study 5 – Kampung Admiralty, Singapore, WOHA Architects

This study presents the example of the Kampung Admiralty, Singapore. Kampung Admiralty was designed to realise a strong vision of an integrated community embracing social activities, leisure and health, a vision which includes connecting with nature in the form of extensive community gardens forming a well-developed biophilic interface focused on the building envelope and the various activities related to modern living. This is the specific subject of the study, but it is also set within the context of Singapore as a leading biophilic city (see biophiliccities.org) where the Kampung Admiralty demonstrates one of the municipality's ideas of 'fusing population density with nature' (biophiliccities.org; see entry for partner cities Singapore). The aim of the study is to showcase the development, but also to demonstrate its attributes as a precedent of integrating nature in a way that produces high biodiversity and routine contact with nature via a functional landscape, thus realising socio-ecological synergies, the principles of which might be followed in a range of development situations.

Various websites provide information on the development, largely based on the social aspects, but particularly impressive is the way that nature is fully embraced as part of the concept. Over half of the building surface area is devoted to nature-inspired green infrastructure, which demonstrates high biodiversity and high performance as a unique socio-ecological landscape system intimately connected with the lives of residents and visitors. Information here is presented through images and information supplied by WOHA, but summary information is available on the WOHA website (https://woha.net/project/kampung-admiralty).

> Kampung Admiralty has the distinction of being Singapore's first integrated vertical kampung. A one stop hub with housing for the elderly and co-located facilities for residents of all ages under one roof. Kampung Admiralty was designed for ease of maintenance and built with sustainable design features. It won World Building of the Year at the 2018 World Architecture Festival and has also received a number of accolades for its innovative approach.
>
> *(www.csc.gov.sg/articles/kampung-admiralty)*

Singapore as a Biophilic City

Singapore features as an impressive example of a biophilic city (see biophiliccities.org) where its structure, identity and international profile are underpinned by biophilic thinking that illustrates high standards and strong visions, having advanced from calling itself 'Singapore Garden City' to 'Singapore: City in a Garden' (Box 9.5.1). Singapore, through changes to its municipal programmes, actions and ideology, including design policies which embrace biophilic thinking, has pursued its strong vision as a nature-filled city with a sophisticated green network,

BOX 9.5.1 KEY FEATURES OF SINGAPORE AS A BIOPHILIC CITY

- Singapore has a substantial and sophisticated green network of paths and trails; these are interconnecting and allow users to walk, jog and cycle within a totally green environment.
- National parks include four nature reserves, more than 300 parks, extensive tree-scape and roadside planting.
- The Garden City Fund was established by National Parks and managed as an independent charity towards optimising green spaces, supporting urban biodiversity, engaging community and enhancing competencies of the landscape industry in Singapore.
- Singapore Botanic Gardens are 82-hectare gardens developed over more than 150 years, with an important role as a leading tropical botanical institute and designated as a UNESCO World Heritage Site in 2015.

- The Centre for Urban Greening and Ecology provides professional skill training at all levels, including Landscape Workforce Skills Qualification, serves as a repository of best practice, and plays a role in advancing urban and green living environments.
- Gardens by the Bay forms a model for sustainable development and conservation and provides an enthralling experience.
- Integration of nature into vertical spaces includes award-winning examples.

(Compiled from biophiliccities.org Partner cities, Singapore accessed 02–03–2022. For full details see original source.)

and has actually achieved an increase in the percentage of the island devoted to 'green' (despite a two million-person rise in population since 1986). Amongst its biophilic attributes, it celebrates the outstanding Gardens by the Bay, featuring 'Supertrees' (Figure 9.5.1), as well as its

FIGURE 9.5.1 Singapore Gardens by the Bay: Supertrees, Flower Dome and Cloud Forest.

Source: (IMG4708 licensed under the Creative Commons Attribution-Sharealike 4.0 international; URL:https://commons.wikipedia.org/wiki/user:Kuroczynski)

internationally famous botanical gardens (see also biophiliccities.org). The experience of Singapore, in taking a strong approach endorses that an ecocentric/socio-ecological focus, reaps substantial dividends and is the proper trajectory for future urban environments.

The Kampung Admiralty

The key features of the Kampung Admiralty are summarised below and explored via a series of images showing several views of and within the building complex. The vegetation structure is impressive, consisting of various planting regimes established on different levels, including ground level. When experienced from within the building complex, the whole gives a strong impression of being surrounded by nature (Figure 9.5.2). From a user perspective, the different levels offer good enclosure, near and distant views and form the context for social and community use. Vegetation systems embrace a combination of greening typologies in a series of terraces, green walls, including sky gardens, a community farm as well as a water feature and surrounding greenery at ground level (Figures 9.5.2 & 9.5.3). Together, they create

FIGURE 9.5.2 Kampung Admiralty: general view showing the well-vegetated terraces.

Source: WOHA, image credit Patrick Bingham-Hall.

FIGURE 9.5.3 Kampung Admiralty: inside the building envelope, views across the terraces from the perspective of the user.

Source: WOHA, image credit Patrick Bingham-Hall.

FIGURE 9.5.4 Kampung Admiralty community garden: family space, places for meeting, experiencing and learning about nature.

Source: bioSEA, copyright: Dr. Anuj Jain.

BOX 9.5.2 KAMPUNG ADMIRALTY HIGHLIGHTS AND PRECEDENT FEATURES

- Vision of an integrated community with one-stop facilities and services, embracing social activities, leisure and health.
- Connecting with nature in the form of extensive community gardens, forming a well-developed biophilic interface within the building envelope.
- Total green areas of 4730 m^2, 53% of the total building plot area of 8980 m^2.
- Audited according to structure based on the vegetation density and characteristics.
- Ground vegetation (level 1), including the water feature and thoroughfare vegetation, consisting of a mix of shrubs and trees based on native vegetation systems.
- Multi-layered roof gardens, levels 3–9, multi-layered plantings consisting of a mix of shrubs and small trees.
- Shrubs, patches of vegetation, levels 2, 6, 7, 8 and 9, dominated by grasses or shrubs with no trees present.
- Green walls, levels 2, 3, 4, 6, 7, 8, 9.
- Community farm.
- Ecosystem services, reference habitat of pristine rain forest as a base comparison using KA's green footprint.
- High biodiversity, due to a diverse planting palette, with a rich mix of shrubs and trees providing fruits, nectar and host plants.
- Water features provide additional habitat, diversity and abundance of insects (higher than surrounding areas).
- Public support which emphasises the value given by the integration of greenery with convenience (especially the Hawker Centre).
- The majority of social survey respondents recognised the importance of biodiversity in a green building.
- Visible biodiversity serves as a positive reinforcement for people who value biodiversity.
- A need to educate some members of the public regarding the value of biodiversity.
- Vegetation types with the most and biggest trees provide higher ecosystem services.
- Linkage with existing nearby areas of natural and other vegetation.
- Biodiversity projections based on KA's current data suggest a substantial increase in species diversity linked to an increase in the area of vegetation.

(*Source*: Dr Anuj Jain bioSEA; information supplied by WOHA (2020), pers.com.

both semi-natural habitat and realise social objectives. As a green complex, it links to nearby nature, including managed parkland and native woodland.

Material supplied by WOHA, which included an independent social and biodiversity report, examined the different vegetation structures represented at the various levels, including assessments of species diversity (birds, mammals, amphibians and reptiles, as well as insects and invertebrates, examining such aspects as feeding behaviour), plus social surveys and ecosystem service benefits. It concluded that 'It is not only a valuable habitat for biodiversity but is highly appreciated by visitors and residents. In addition, it provides significant ecosystem service benefits'. For example, in assessing how residents and visitors viewed the development, independent survey indications were that the combination of various plantings together with facilities was highly effective and appreciated for their biodiversity, contact with nature and convenience.

Conclusions

Singapore is one of the most impressive biophilic cities, where its modern identity is underpinned by biophilic thinking and the reality of initiatives that are recognised as setting high standards for the integration of nature throughout the city. The example presented in the Kampung Admiralty extends the vision and practice concerning nature, built forms and social enterprise, and demonstrates practice which sees the building as a self-sufficient socio-ecological complex which places people in nature but also seeks to address concerns regarding biodiversity. By adopting an approach which utilises vertical space and terracing, it is possible to see how the design uses the surface area of the building to advantage to maximise planting, and similarly maximise interaction visually and physically, using nature as a mechanism and setting to achieve multi-outputs, the effectiveness of which is confirmed by independent audit. As such, it forms an important and impressive precedent for the realisation of effective biophilic-focused urban forms.

Reference

WOHA. (2020). *Kampung Admiralty Biodiversity, Social & Ecosystem Services Audit,* submitted 2 October 2018. By Dr. Anuj Jain BioSEA LLP, supplied by WOHA, 29 HongKong Street, Singapore.

9.6 Case Study 6 – Forest School: Establishing Environmental Literacy and Environmentally Focused Life Skills

This case study concerns ideas of developing environmental literacy in young children as a fundamental aspect of education, considering the Forest School initiative. As such, it stresses the importance of children having the skills, knowledge and confidence to benefit from accessing the natural world, and takes account of the fact that the development of environmental literacy, the capacity to understand and experience nature firsthand, is a fundamental life skill that needs to be supported in children. These are skills that are essential for later life regarding an individual's ability to access the positive restorative state that can be experienced through nature and related circumstances, as well as developing appropriate attitudes towards the environment. The featured forest school example is Our Lady Queen of Peace Catholic Primary School (ourlady.worcs.sch.uk), which has been selected on the basis that it features quite a modest wooded area within the school grounds and shows what can be achieved with strong and enthusiastic leadership.

The 'Forest School' movement addresses the fact that children need routine exposure to the natural world, and that such exposure provides important positive outcomes concerning educational achievement and social skills, but also helps to break the cycle of Generational Environmental Amnesia by stimulating the child's innate biophilic self. In particular, the Forest School approach emphasises the role of schools as influencers in promoting healthy environmental behaviours and responsibilities that are likely to persist through into later life.

> Schools as influencers – making outdoor learning and play part of every child's life. Schools, by sending the clear message that getting outdoors is important to children's wellbeing and development, can help make outdoor learning and play become part of every child's everyday life.
>
> *(Muddy Hands Report Prisk & Unsworth, 2018)*

Forest School was developed from Scandinavian experience and spread to the UK, the USA and Canada, with similar initiatives in Australia. In the UK, the movement is underpinned by the Forest School Association, who provide guidance, promote training and uphold standards. Similarly, other countries emphasise environmental literacy as a critical aspect of children's experience within the formal educational process and advocate the adoption of Forest School-type approaches with schools and other partnerships which may be extended to address different public agencies. All these organisations provide online resources, professional support and publications. For example, in the USA, the movement is well developed through the Forest School Foundation (www.theforestschoolfoundation.org) and Green Schoolyards (greenschoolyards.org/greenschoolyardsamerica) and in Australia, the Australian Forest School Association (https://australianforestschoolassociation.org).

The Forest School programme has a set of principles which were agreed upon by the UK Forest School community in 2011. These recommend a range of learner-centred processes by which to create a community for development and learning focused around woodland-based activities. Guidelines are comprehensive and emphasise being responsive to the needs and interests of learners, requiring careful planning, appropriate dialogue and relationship building. The principles embrace play and choice as integral parts of the learning process and in providing a stimulus for all learning preferences and dispositions, for example, involving reflexive practice, 'scaffolding' and tailoring experiences to learning (for full details, guidelines and principles, see The Forest School Association, www.forestschoolassociation.org).

Whilst woodland is the indicated environment for a Forest School, the Forest School Association encourages schools which might only have a few trees to develop a Forest School plot, the principle being that a woodland type of environment, even a small area of land, is well suited to match the needs of the programme and learners. The underpinning requirement is the provision of space and an environment which offers opportunities to explore and discover.

Forest School at Our Lady Queen of Peace Catholic Primary School (OLQP)

The school caters for primary education, including early years foundation stage from ages four to 11, and although surrounded by built forms and having only a limited amount of natural space, shows what can be achieved. The forest school area consists of a group of mixed trees which

FIGURE 9.6.1 Forest School at Our Lady Queen of Peace (OLQP). General view inside the site, notice the log circle beyond the large yew tree.

forms quite a feature of green in the area, situated alongside a road. Although small, the site is well suited to the aims of the programme and accommodates small groups of children quite comfortably, providing a space that offers 'opportunities to explore and discover' through contact with nature.

The OLQP Forest School site encompasses:

- A small grove of about six closely spaced mature trees with various sturdy shrubs as understory, essentially forming a wooded area but with no threatening characteristics.
- Secure with no fear of getting lost, providing easy surveillance of the children and ground conditions accessible in all weathers.
- Good enclosure and atmosphere as a 'special' place.
- Provides an effective transition into nature, emphasised by the need to pass through a purpose-built wooden gate inscribed with the words 'Forest School OLQP'.
- Being within the school grounds benefits from wider school security.

- Good opportunities for exploration and discovery, including hiding and associated play-focused activities.
- A fire pit and log circle which focuses on gathering together, shared activities, fire lighting and cooking skills.
- Wooden shed for additional shelter and storage.

FIGURE 9.6.2 Forest School at Our Lady Queen of Peace (OLQP): the nicely inscribed entrance gate off the main playground, which also emphasises that this is a special place to be enjoyed but which also requires care and looking after.

- Activities supported through a trained Forest School leader, assisted by volunteers, providing a good adult-to-pupil ratio.
- A forest school curriculum that outlines skills, challenges and outcomes across years.
- Woodland-based activities, scope to collect, touch, encounter and construct, enhanced by the importing of additional materials as required.
- Low branches that are easily reached by all the children and which, among other things, offer opportunities for den building.
- A 'mud kitchen'.
- Supported by refreshment breaks, drinking hot chocolate and toasting marshmallows.

The forest school sessions are held regularly, weekly, the children having been kitted out at the start. The session observed began with the children seated on the log circle and the outlining of the rules and responsibilities, followed by different activities including storytelling, finding a hiding place, collecting material to make 'Stickman' – 'make and decorate your own Stickman from natural materials made in the wood' (see the popular children's book by Julia Donaldson, 2016) practising tying knots and finishing with a hot chocolate drink, all the children being fully immersed in the various tasks assisted by the helpers as required. The session is playful and is underpinned by the ethos of playful exploration in nature, clearly enjoyed (see 'Likes', Box 9.6.1) and emphasised as special and memorable by hot chocolate, or in other sessions, toasting marshmallows on the campfire.

BOX 9.6.1 SOME CHILDREN-IDENTIFIED 'LIKES' AT OLQP FOREST SCHOOL

I like to collect sticks to make 'Stickman'.
I love marshmallows in the hot chocolate, and marshmallows on the campfire.
I like playing with my friends.
I like playing make believe games with the grown ups.
I like playing calm games and having lots of space.
I like making things in the mud kitchen.
I like climbing in the trees and den building.

Forest School Skills and Challenges

Forest School takes into account that children experience and learn about nature through play, responding to their natural enthusiasm and the stimulation of activities which have eco-haptic associations, essentially doing things in nature, foraging for materials, exploring the environment and working with others. The curriculum addresses a range of six skill sets which concern using tools, fire lighting and cooking, den building, ropes, living things and craft. Starting with relatively simple tasks, these progress to more independent activities; for example, for the theme of fire lighting and cooking, progressing from 'sitting safely around a campfire' to 'independently light a fire' to 'cook a simple meal to serve to a larger group of friends' – instilling confidence as challenges are set and achieved.

Within the Forest School programme, it is easy to see how nature provides a natural platform or pathway for the development of skills and social activities, and Forest School activities can be considered more

FIGURE 9.6.3 Activity, action and hands-on experience: using tools and craft activities, two of the skill sets emphasised through Forest School activities. Also notice the good cover of leaves on the ground, ideal for collecting.

Source: (Photo Annette Vennard, Our Lady Queen of Peace Catholic School.)

FIGURE 9.6.4 Toasting marshmallows on the log fire: fire lighting and cooking, activities referred to in the 'skills and challenges' and as a 'like' by the children.

Source: (Photo Annette Vennard, Our Lady Queen of Peace Catholic School.)

broadly within a socio-ecological remit. It is perhaps this combination that makes it a particularly important aspect of educational development and relevant to future adult actions.

As Annette Vennard, Forest School leader at OLQP, observes:

> Forest School enables children to grow in confidence at their own rate. They have time and space to achieve things independently. They learn social skills, and become aware that they can help each other. They also begin to learn the consequences of their actions. They only have three rules – to keep themselves safe, their friends safe, and the animals and plants of the wood safe. They learn specific physical skills, such as using tools, like full size saw and hand drills.
>
> They gain a greater knowledge and understanding of the world, sometimes shown in a sense of awe and wonder. My most vivid recall of this is a boy, who lives in a flat in the middle of the city, sitting with a helicopter sycamore seed on his nose. He wasn't engaged in the main activity of the morning, but he looked at

FIGURE 9.6.5 Den building: one of the skills and challenges – 'build a den in a small group with a set of sticks, ropes and tarpaulins'.

Source: Image credits: Annette Vennard

> me and said, 'Isn't it amazing that that huge tree over there came from a tiny seed like this'.
>
> *(Annette Vennard, 2022, pers. Comm)*

> To sum it up one little girl said she liked having fun and playing with her friends. . . . It seems there is something to please everyone at Forest School.
>
> *(Annette Vennard, 2022, pers. Comm)*

Conclusions

The experience at OLQP demonstrates the importance of a Forest School programme in aiding children's development and what can be achieved by regular contact involving the power of nature, even in quite a modest area, establishing attitudes and experiences that are an important part of life skills. The programme at Forest School OLQP emphasises regular and sustained contact with the natural world, developing and building on skills and the importance of playing in nature. As such, they concern moments of intense sensory feedback typically associated with deeper biophilic interaction, involving exploring, making haptic connections and the physicality of the encounter, climbing and swinging on branches, including what appear to be moments of reverie, 'awe and wonder'.

However, the Forest School is also a place where social skills are learnt, nature associations acting as a cultural and social mediator, children being aware that they can help each other, a power which, it is suggested, lasts beyond the initial Forest School activity and is seen to aid the child's progression and skill acquisition in other areas. The children at OLQP have many different backgrounds, but can respond to each other through the common language and communication of biophilia.

Reference

Donaldson, J. (2016). *Stick Man*. New York: Scholastic, Alison Green Books. Prisk and Unsworth. (2018). Muddy_Hands_Global_Report_November_5_pdf

9.7 Case Study 7 – Co-Design, Partnerships and Environmental Stewardship: the Case Study of Gardens for Wildlife Victoria

The issue addressed in this case study concerns the problems surrounding how to move to a position where environmental stewardship becomes the norm and is reflected in actions on the ground. It looks at what processes and partnerships might be involved to reach such a situation by exploring the example of 'Gardens for Wildlife Victoria' (see gardensforwildlifevictoria.com).

> Gardens for Wildlife Victoria is a network of community volunteers and staff in government agencies or departments (primarily local government) that support each other to increase awareness and knowledge in the community about the importance of creating and improving habitat for locally native fauna and flora, and to involve them in doing so.
>
> *(gardensforwildlifevictoria.com [our work])*

FIGURE 9.7.1 Richard and Irene's garden in South- East Melbourne managed for nature conservation and to attract wildlife under the scheme.

Source: Image credits: Richard Kelly

The issues at hand are several, and relate to urban and bringing private land into the nature/wildlife system in a situation where ownership is diverse, but in fact resources, in terms of both the extent of private holdings in the form of gardens and in terms of people, are high. In the example of Gardens for Wildlife Victoria, the strategy developed is part of a wider wildlife conservation aim, but in fact the successes of the venture show a strong socio-ecological context involving habitat improvement and a wider response concerning place identity, as well as social and personal gains (Figure 9.7.1)

The organisation has a comprehensive website, and its operation has been extensively reviewed by Mumaw (see Mumaw & Raymond, 2021; Mumaw & Mata, 2021; Mumaw et al., 2019; 2017; Mumaw, 2017; Mumaw & Bekessy, 2017). For example, as Mumaw and Raymond (2021) observe, moving from a situation where roles have tended to be entrenched to one where shared/common values and trust are agreed in a system of partnerships and a process of co-design, based on:

- *Common values and trust,*
- *Focusing on problems differently,*
- *Not being bound by traditional roles,*
- *Empowerment of individual actors,*
- *Encompassing a sense of intrinsic motivation embodied in caring for nature/the environment,*
- *Strong place identity,*
- *And visible action on the ground.*

The Programme

Actions are focused on place and local place identity, and recognise that wildlife conservation has in the past failed to embrace the potential of urban private gardens as a wildlife resource, and that existing urban dynamics and city management regimes can cause difficulties in encouraging a sense of stewardship. Intrinsically, the scheme takes an ecocentric approach and recognises the potential of urban and 'what could be'. Accordingly, it progresses the situation towards one where the urban environment in ecological and socio-ecological terms moves to a closer definition of being 'healthy'.

This chief mechanism is through private-public-civic collaboration, which clearly can provide a powerful and effective means of fostering (developing and expanding) stewardship programmes and meeting ecological needs, coupled with a range of community and individually related outputs concerning well-being. As the Gardens for Wildlife Victoria website states:

> We promote community-local government partnerships as hubs to develop wildlife gardening programmes that meet their social and ecological needs and aspirations. We foster collaboration, connections and wellbeing alongside environmental objectives.'
>
> *(gardensforwildlifevictoria.com)*

Mumaw and Raymond (2021) reviewed the scheme in some detail, analysing the approach and its success. They emphasise that embedding nature stewardship in everyday practice requires landscape

FIGURE 9.7.2 Gardens for Wildlife Victoria logo.
Image credits: Laura Mumaw (on behalf of GWV)

approaches with supportive forms of governance, policies, and institutional settings involving collaborative multi-actor networks that span spatial and political boundaries and actors that include governmental entities, businesses, the third sector (voluntary, not for profit associations) and citizens, involving households, families and individuals.

Gardens for Wildlife Victoria challenges norms and cultural perceptions, as well as the inertia inherent in environmental generational amnesia, including the divide between urban and natural. Thus, while conservation activities in wilder places form an important aspect of leisure activities action is not usually transferred in any co-ordinated way to the wider urban environment. Under the

BOX 9.7.1 THE FEATURES OF GARDENS FOR WILDLIFE VICTORIA, SUMMARISED FROM MUMAW AND RAYMOND (2021)

- Gardens for Wildlife Victoria is a rapidly growing network of civil and agency actors that seeks to embed biodiversity stewardship across landscapes in the state of Victoria to support human and environmental health.
- Facilitates the development of civil-agency collaborations to design and execute programs that engage local residents and businesses in wildlife gardening to complement biodiversity stewardship on public land, for long-term social and ecological health.
- The network links programmes across different local government areas and seeks to incorporate their work into government policies and activities.
- The programme was launched in August 2016 as a 'governance transformation initiative' arising from an enquiry into urban biodiversity governance in Victoria.

- Its purpose was to find a way for biodiversity conservation to be understood and embedded in the lives of all Victorians, particularly in cities.
- Core principles include co-design and management of wildlife gardening programs by citizens working directly with agency staff; localised design of programs supported by working models that can be adapted to local contexts; social connection and well-being alongside biodiversity objectives; inclusion of all interested citizens with support, including face-to-face visits in participants' gardens; and evaluation and strengthening the impacts of its activities.
- The focus is on urban landscapes – referring to landscapes in cities, townships and villages in rural or city areas; privately managed plots of land, particularly residential properties, which historically have been omitted from state biodiversity stewardship strategies; actions include nurturing locally native flora and fauna, and ecological diversity.
- Gardens for Wildlife Victoria provides program development guidance, mentoring, support and advocacy through events, workshops, training sessions, presentations and government liaison. Participants join the network through a snowballing process, whereby involved actors invite others or interested parties make contact with the network.
- In a little over three years after the launch in late 2019, the network included over 300 members and continues to grow.

(Based on Mumaw and Raymond, 2021; for full details and in making further reference, consult the original source.)

leadership of the programme, municipalities have been encouraged to foster biodiversity stewardship collaborations with civil society and active citizen groups.

The responses to the programme analysed by Mumaw and Raymond (2021) sit well within the idea of a socio-ecological perspective and a re-awakening of purpose, a sense of well-being, a renewed passion for wildlife as well as the exchange of knowledge and ideas, which boosted capability, confidence and the implementation of initiatives, for example, in helping participants overcome challenges, through handholding and bouncing or building off ideas.

BOX 9.7.2 GARDENS FOR WILDLIFE VICTORIA: HIGHLIGHTS

Goals

- To nurture our flora, fauna and environment.
- To build knowledge, skills, confidence and ownership for caring for the land and its wildlife.
- To strengthen connections between residents, nature, community and place.
- To improve well-being through experiencing and caring for nature.
- To build a network of environmental stewards and champions.
- To collect and share stories and data about the impact of our work.

Actions

- Promote community-local government partnerships as hubs.
- Develop wildlife gardening programmes that meet social and ecological needs and aspirations.
- Foster collaborations.
- Share ideas and resources.
- Educate.
- Celebrate successes.
- Strong web platform.
- Case study videos.
- Link individuals.
- Provide workshops and guidance.

Features

- Network with a vision – of wildlife gardening as caring for community and nature.
- Address the everyday – relevant to everyday lives and the places where individuals live, meaningful for each person.
- Connecting – practical ways of connecting people to their environment and each other involving a common purpose.

- Place and ecological identity – focus on 'your' environment, the wildlife where you are, the things that you can do yourself, a feeling of belonging.
- Health and well-being – positive for personal and emotional health, social health, ability to be resilient, moving towards ecological health.
- Sense of achievement – benefit and results can be seen multiplying through neighbourhoods.

Processes: a process where civil and agency actors are recognised to have equivalent contribution rights and responsibilities.

- Leaders and champions – the presence of leaders and champions, involving collaboration and understanding with a clear shared purpose and responsibilities.
- Flexible timescales – recognise that co-design takes time and may not accord with conventional local authority financial/budget patterns or typical timelines.
- Empowerment and empathy – allow the community to lead and set the pace.
- Open to ideas – responsive to alternative ways of doing things.

Notes

1 The goals are as stated by Gardens for Wildlife Victoria on their website. gardensforwildlifevictoria.com
2 Other entries are extracted from details given by Mumaw and Raymond 2021 in their analysis of the scheme and gardensforwildlifevictoria.com, selecting key words and descriptions that best highlight actions and features.
3 Any further reference should be made by consulting the original source.

Conclusions

The significance of the Gardens for Wildlife Victoria is considerable, in that it places landscape and wildlife stewardship within a different paradigm of shared environmental responsibility and understanding.

FIGURE 9.7.3 Nature stewardship workshop.

Source: Image credits: Laura Mumaw,

In addition to mobilising additional resources, land and labour, being founded upon notions of stewardship, it identifies a powerful niche of operation and enthusiasm which is particularly relevant to urban circumstances. Nature here is both the driver and beneficiary, in that the human-nature bond and biophilia underpin its action to create a strong positive feedback system which satisfies our need for contact with nature in ways that are personal to the individual, i.e. in which cultural and biophilic identities are readily defined and expressed. As Laura Mumaw observes:

> Acts of gardening considered mundane can foster indigenous flora and fauna in backyards and across urban landscapes through interconnected social and ecological mechanisms. Wildlife gardening can deepen and expand nature stewardship by residents while providing personal wellbeing benefits and strengthening community connections. Pointing to codesigned wildlife gardening program successes in diverse municipalities, we encourage local authorities to collaborate with community members to codevelop

> wildlife gardening programs that strive to achieve municipality-wide well-being and biodiversity objectives. We recommend networking between programs to foster shared learning, empowerment, and innovative new approaches.
>
> *(Mumaw 2022, pers. com.)*

Programmes such as this define the way forward in developing approaches which are sustainable and powerful, and appear to satisfy a hunger for biophilic fulfilment. These programmes demonstrate how our relationship with nature and the natural world is being redefined in terms of opportunities, responsibilities and circumstances, and how the human-nature bond readily re-surfaces when prompted as a powerful motivator. Again, as Laura Mumaw considers:

> Wildlife gardening programs across the world, adapted to the land and cultures in which they are situated, can contribute to the achievement of global sustainable development goals, including those related to urban sustainability, human well-being, and conservation of biodiversity.
>
> *(Mumaw 2022, pers. com.)*

References

Mumaw, L. (2017). Transforming urban gardeners into land stewards, *Journal of Environmental Psychology* 52, 92–103.

Mumaw, L., Bekessy, S. (2017). Wildlife gardening for collaborative public – Private biodiversity conservation, *Australasian Journal of Environmental Management* 24(3), 242–260.

Mumaw, L., Mata, L. (2021). Wildlife gardening: an urban nexus of social and ecological relationships, *EcoEvoRxiv Preprint*. https://doi.org/10.32942/osf.io/9rkhm

Mumaw, L. M., Maller, C., Bekessy, S. (2017). Strengthening wellbeing in urban communities through wildlife gardening, *Cities and the Environment* 10(1), 6.

Mumaw, L. M., Maller, C., Bekessy, S. (2019). Assessing and strengthening community capacity building in urban biodiversity conservation programs, *Cities and the Environment* 12(2), 4.

Mumaw, L. M., Raymond, C. M. (2021). A framework for catalyzing the rapid scaling of urban biodiversity stewardship programs, *Journal of Environmental Management* 292, 112745. https://doi.org/10.1016/j.jenvman.2021.112745

9.8 Case Study 8 – Biophilic Workplaces – The Case of Amazon Spheres, Seattle, USA (with the contribution of Matthew Wood, Principal at *SiteWorkshop, Landscape Architecture*, Seattle)

This case study concerns innovation in the development of the office environment, exploring the ways that office space can be re-configured essentially as a system, in which nature and biophilic interaction provides the context in producing an environment for thriving and supporting well-being at work. This is explored through the case study of Amazon Spheres, which was completed in 2018 and features a highly innovative office environment which, in concept and reality, suggests that the way that we conceptualise work spaces needs to be challenged, but also has an impact on our understanding of how we experience nature – where, when and how – including the situations and circumstances by which it is made available or presented.

In approach and philosophy, Amazon Spheres challenges the notion of what a workplace could be, and extends our understanding of how we can place ourselves in nature to significant advantage but also, in doing so, aim to help conserve and maintain plant diversity and understand more fully our responsibilities to the natural world.

Amazon Spheres, through one of its designers, Matt Wood, takes

> a living systems approach to integrate nature into the built environment . . . to creatively develop immersive, nature-rich landscapes that benefit both people and wildlife.
>
> (*Wood, 2022*, pers. comm.)

Accordingly, Amazon spheres seeks to realise the benefits of being in a nature-rich environment and emphasise how a biodiverse plant ecosystem can flourish and encourage positive connections with nature.

> The Spheres are a result of innovative thinking about the character of a workplace and an extended conversation about what is typically missing from urban offices – a direct link to nature.
>
> (*www.seattlespheres.com/*)

The characteristics of the approach take inspiration from the deeper levels of engagement and positivity that can be obtained through biophilic situations, and make a distinction between the adding of plants to existing office spaces as superficial ornamentation with limited biophilic effects, and the impact of being immersed in planting displays where the plants are central to the experience; they are intended to be seen as living beings, and provide the multisensory and restorative experiences associated with the power of nature.

The Spheres as an Alternative Work Experience

The Spheres were established as an alternative workspace to the traditional office environment by providing Amazon employees and visitors opportunities to immerse and interact with nature throughout the year (Figure 9.8.1). The concept is underpinned by the idea that spaces characterised by having significant biophilic elements can improve health, brain function and inspire creativity, to realise an inspiring design that is supported by biophilia at its core. It is described as free-flow workspace with a range of spaces for people to sit with their laptops or simply to walk along during a meeting.

> The original inspiration for the Spheres was, quite explicitly, an iconic botanical conservatory. In retrospect, that touchstone

FIGURE 9.8.1 View that captures life in the Amazon Spheres, with users immersed in a nature-rich environment.

Source: Image credits: Stuart Isett.

> established a fundamentally consequential direction for the project, allowing the team to push beyond conventional approaches to interior landscapes and biophilic design.
>
> (*Wood, 2022*, pers comm.)

> [It features] a variety of informal gathering areas, food service and events spaces integrated throughout the building and interior landscape, the Spheres is a place for employees to connect with each other. At the same time, a variety of immersive planting displays engage employees in the natural world, creating a restorative environment that relieves stress, invites curiosity and sparks creativity.
>
> (*www.siteworkshop.net/*)

> We need to start by emphasizing the bio- in biophilia, and hopefully that opens up some new design opportunities that lead us away from any tired conventions that prevent us from actually seeing the plants as plants, or even noticing them at all.
>
> (*Wood, 2022, pers. comm.*)

The Spheres host more than 40,000 plants, with around 25,000 in the green wall, in which collections impersonate the lush and bold foliage of rainforest environments from Central and South America and tropical Asia, such as China and Japan. These translate into vertical gardens of the Canyon Living Wall (Figure 9.8.2) to the flora-filled Fernery, with additional elements including fish tanks. The microclimate, whilst highly controlled, is characterised by temperatures similar to a conventional office environment (kept at around 22 degrees Celsius) but a higher humidity (kept at around 67% average); beneficial insects are released to manage the pests (www.seattlespheres.com/the-plants).

The nature of the project required an integrated approach which meant that the design had to consider not only the planting palette, but also the building's environmental systems, informing decisions about glazing, lighting, interior finishes and materials to ensure that the building environment supported a thriving plant collection, while also remaining comfortable for human co-occupants. Such understanding supported initial lessons that informed Wood and his team's approach to the Spheres, with the key requirement that design teams need to appreciate plants as living things so users can draw benefits from their

FIGURE 9.8.2 The visual impact of the Green Wall at Amazon Spheres arises from challenging the relationship between technology and horticulture. It holds 25000 individual plants, including around 200 different species.

Source: Image credits: Stuart Isett.

own vigour, a situation which translates into the challenge of exploring the relationship and interface between a botanical conservatory and an office environment.

Matt's observations arise from seeing sad displays of plants in continuous decline where the microclimate conditions, light and space have been underestimated. As he explains:

> Don't rely on 'common houseplants' any more than necessary. Of course, there are very good reasons to use 'common houseplants'. For one, they are proven to be reasonably reliable in low-light, interior conditions. And significantly, they are available in the trade, so we can be fairly confident that we can procure them and in quantity. However, there's an element to which their ubiquity contributes to their invisibility. Admittedly, this is all relative. What's boringly familiar to one person, may be quite exotic to another. And then chasing novelty is a bit pointless, especially in this Instagram-driven age of consumerism. Today's Geogenanthus is tomorrow's Pothos. So the challenge becomes presenting

plants in a way that invites attention, no matter how familiar they may be. It's no small feat to accomplish that with any degree of grace. [The challenge is to] recontextualize them.

Provide conditions that allow plants to truly thrive, not merely survive. No matter what you're trying to grow, plants that aren't healthy offer little biophilic value . . . In fact, they just bum people out. It seems so basic, but providing a horticulturally appropriate environment is far-and-away the biggest hurdle to successful interior landscapes. It's gotten to the point that when we're approached by architects about interior projects, I first try to talk them out of it . . . putting plants in conventionally designed office buildings is a terrible idea; it's such a hostile environment, too dry and far too dark . . . it's our responsibility to provide proper life support systems, starting with appropriate lighting and continuing through skilled maintenance.

(*Wood, 2022,* pers. Comm.)

Focusing on Biophilia and Its Effects

There is positive relationship between incorporating biophilic elements and visually stimulating spaces and well-being in the workplace, including a reduction in stress levels (Lottrup et al., 2013; Yin, 2020), positive impact on happiness and reduction in the number of absences and sick leave days (Elzeyadi, 2011). Other studies have also shown an increase in engagement and productivity and thus financial gains (Browning et al., 2015). These are, in turn, supported by research on Attention Restoration Theory (ART), in the sense that a recovery from mental fatigue associated with work events (involving the quality of fascination and sense of being away) will promote better focus, attention, engagement, creativity and productivity.

- Fascination – Using plants that are likely unfamiliar to most allowed us to capture people's attention not just with their novelty, but also with some botanically interesting morphology, relationships, colors, textures, scents. And then the planting design is deliberately maximalist. In part that helped us integrate a truly diverse collection, but we also needed to keep things interesting for people who might visit a couple times a week. Hopefully they could see something new each time. Monocultural blocks that are comprehended at highway speed just wouldn't cut it at this scale. There's plenty

> of repetition, but also an abundance of diversity throughout to the point where we hope apprehension exceeds comprehension. . . . [and that the] more casual visitor experiences it differently.
>
> - Being away – The sense of being transported starts at the entry where people are greeted with a distinct, earthy odor, a fecund scent that lets visitors know they are somewhere special. We also really worked hard to create a sense of immersion, but the planting area is limited to something like 4,300 sf each in the north and south sphere. Not much. So we played with plant height and density, along with speed of circulation, to create the illusion of much more extensive planting. It can be oddly disorienting. I've given a number of tours where people lose track of where they are, thinking there's an entirely separate sphere to go into next (all three spheres share a common volume and environment). Or they forget entirely that they are in the middle of downtown Seattle, despite the surrounding towers being visible through the glass.
> - Extension – We integrated a variety of circulation routes and vistas throughout the building that allow the same planting areas to be experienced from very different perspectives. There are also a range of microclimates that suggest a wild landscape of sorts, a living system, as well as a stream that appears to cascade over a waterfall at the entry. And then there are numerous botanical stories woven throughout, relationships between plants, ethnobotanical uses, conservation stories, etc.
>
> *(Wood, 2022, pers.comm.)*

The building itself is a unique landmark in Seattle, with the design being driven by the plants' requirements, being mostly built of glass to maximise natural light in supporting the ecosystem. The spatial layout of the building allows for in-depth exploration of the planting displays (Figure 9.8.3), for example, the open staircases provide a view of the green wall as people walk through. It encourages people to be curious, learn and think differently in this environment, restoring cognitive functioning, but also to feel refreshed, energised, losing oneself and by being in close proximity with nature. However, as Matt Wood explains:

> Now this raises a more general question: if we go to the effort and expense of creating a horticulturally appropriate environment inside an office building, what other opportunities can we capitalize on? If we're going to allocate energy for horticultural lighting

FIGURE 9.8.3 The layout of paths and spaces allow sensory interacting with the diversity and lushness of planting displays, from olfactory to tactile to visual stimulations.

Source: Image credits: Stuart Isett

> and water for irrigation and fog, can we make that an investment in a greater good? For this project, we chose to prioritize ex-situ conservation. Why limit ourselves to common houseplants available at every big box store, when we can create an interior environment found in only a few botanical gardens that allows us to preserve species that are unknown or threatened in the wild?
>
> *(Wood, 2022, pers.com)*

Conclusions

Amazon Spheres, as in common with other case study examples, both challenges and extends our concept of nature in our designed world. In re-defining the workplace, it emphasises the understanding that nature is something that has to be experienced as an integral part of living, not as something set aside. As such, it involves the re-conceptualisation of the office as a biophilic space in the form of a real ecosystem, in which work-based activities are supported by the therapeutic effects of nature, as good places to be.

Amazon Spheres challenges notions of how we experience the natural world in our everyday lives, and further breaks down some of the barriers regarding different space and building typologies, in that biophilic solutions offer particular qualities regarding how we can and need to embrace the power and aesthetics of nature and how we dwell in nature in the modern world. Despite relying heavily on stewardship, it applies ideas relating to the restorative effects of natural environments, and seeks to provide high-quality conditions for human habitation – specifically in work situations, to take account of the stress and mental exhaustion that can be involved. As a response, it is informed by theory and translates best practice into new situations, which can, perhaps, be considered part of the evolutionary process of understanding more fully our relationship with nature, its power and our responsibilities towards the environment, while at the same time designing places which are dramatic, inspiring and environmentally beneficial.

References

Browning, B., Garvin, C. Ryan, C., Kallianpurkar, N., Labruto, L., Watson, S., Knop, T. (2015). *The Economics of Biophilia*, 1st edition (Why designing with nature in mind makes financial sense). Terrapin Bright Green. www.terrapinbrightgreen.com/wp-content/uploads/2012/06/Economics-of-Biophilia_Terrapin_2015p.pdf

Elzeyadi, I. (2011). Daylighting-bias and biophilia: Quantifying the impacts of daylight on occupants health. In *Thought and Leadership in Green Buildings Research* (Greenbuild 2011 Proceedings). Washington, DC: USGBC Press.

Lottrup, L., Grahn, P., Stigsdotter, U. K. (2013). Workplace greenery and perceived level of stress: Benefits of access to a green outdoor environment at the workplace, *Landscape and Urban Planning* 110(1), 5–11. https://doi.org/10.1016/J.LANDURBPLAN.2012.09.002

Wood, M. (2022). *Email Exchange with Matthew Wood*, 26 June.

Yin, J., Yuan, J., Arfaei, N., Catalano, P. J., Allen, J. G., Spengler, J. D. (2020). Effects of biophilic indoor environment on stress and anxiety recovery: A between-subjects experiment in virtual reality, *Environment International* 136, 105427. https://doi.org/10.1016/J.ENVINT.2019.105427

10

A BIOPHILICALLY FOCUSED ENVIRONMENTAL AESTHETIC

Introduction

The ways that we interact with urban landscapes and the consequences of such interaction are receiving increasing attention under a number of headings and disciplinary remits. By examining different ideas and looking at a range of examples which consider different aspects of interaction, and in particular the connections that are indicated in biophilia, i.e. expressing the nature-human bond, a range of theoretical perspectives have been identified and discussed which describe how the urban landscape system functions and performs in relation to the human condition. From these, it is possible to understand more explicitly the development and management of the urban environment as a unique ecosystem which reflects human need, as well as broader environmental priorities.

These perspectives concern the different disciplinary settings which underpin their development and address, or must address, the reality of urban living, but are of course set within the context of the everyday dynamics, ecological and social properties of the places involved. As we have stressed, it is this reality, the everyday experiences of individuals as they move throughout the environment, that is important and represents key aspects of city identity and people-place relationships.

These findings are, in turn, contained within broader ideas of urban environments as functioning ecosystems which, although not necessarily conforming to natural systems, embrace the power of nature. As such, they can and must be studied in their own right, encompassing the range of their diversity and understood within the wider remit of sustainability. Considered together, they also reflect the development of an environmental urban aesthetic that embraces this same power. It is the parameters surrounding this aesthetic that are discussed in this final chapter, and concern nature as an integral part of urban life and lifestyles, as the benefits are

DOI: 10.4324/9781003099758-11

recognised and the multiple ways that it can be experienced across the full range of urban situations understood.

In the context of the above, writing this book and exploring the various facets of information can be described as a journey, involving a re-examination of the results of the varied research enquiries in which we have been involved over the years, but also more literally as we have walked various local routes close to the places where we live. Having explored theory, presented ideas and collated selected case studies, these theoretical perspectives have become the reality of our own experience, understanding the ways that, as individuals, we respond to different environmental encounters in the routine of our everyday activities to an extent that one finds oneself immersed in a variety of situations where it is possible to see, hear and feel the 'ecological hum' and the positive biophilic influences that such places provide.

This concerns how urban nature can be present and appreciated and is waiting to enrich and support our lifestyles in all sorts of positive ways and in diverse situations, even though there exist substantial discrepancies, unrealised potential and mixed attitudes regarding opportunities to experience such relationships. Many ideas explored are underpinned by practice, and we have become increasingly sensitive to the issues and opportunities present in developing urban landscape systems, including the actions and ideas that are being promoted and achieved in all sorts of locations, how they address and respond to different urban contexts and drivers. One of our aims has been to promote a similar sensitivity in others both in visioning 'what could be' and achieving the same.

Holistic Thinking

Understanding of the urban landscape essentially requires an all-embracing approach which encompasses the reality of the human experience, and accounts for and is informed by the various perspectives and examples explored in this book. While the situation can be considered complex, as material is drawn from a range of disciplines, in reality, it is informed by these same disciplinary remits. Ideas are thus multi-faceted, or multidimensional; they dwell across the different disciplines rather than being bounded by them, embracing, for example, social, cultural, health, environmental and design perspectives. It is within a broader holistic understanding of the urban environment and environmental interaction that these different perspectives can inform and be applied in terms of practical management and improved provision. It is a holistic ecocentric stance which takes us to a position where multi-outputs are derived from coordinated efforts, and by combining many small actions the overall result can be considerable.

This also means that the activities or actions in one niche, e.g. by a private company, an individual, a local authority or charity, are all significant when counted together if viewed as parts of a wider, all-embracing action, a linked system, underpinned by taking an ecocentric approach which brings actions together under a common purpose, understanding the nature, needs and potential of the urban landscape and wider urban environment.

The undertow of environmental enthusiasm (involving an ecocentric stance and a biophilic frame of mind) is present across sectors, and is evident in looking at the case studies with substantial commonality regarding ideas, approaches and enthusiasm and in embracing an urban order which allows nature to take the lead. This is set within the notion of a socio-ecological relationship (Frank et al., 2017) which concerns the fabric of urban places, and the ways that people interface with them and are encouraged or permitted to interface with them (as opposed to being restricted in doing so) plus understanding of the way that the body and the mind experiences and orientates itself when moving through or relaxing in those same environments. This includes understanding the conditions that need to be met to make these experiences fulfilling, mentally, physically, socially and culturally.

Broadly speaking, in a holistic framework, several concepts can be highlighted which arise from enquiry, and in which the wide range of theoretical and practical aspects sit quite comfortably:

- **The urban environment as a socio-ecological system:** relating to the properties of the environment and its potential performance as an ecosystem which satisfies multi-outputs particularly within ecological, social, community, cultural and health remits.
- **The choreography of the landscape experience:** involving and demonstrating the experience of the individual as they move through or engage with the environment and the way that the mind and body are orientated and influenced by different environmental parameters.
- **A continously connected biophilic landscape:** concerning the distribution of biophilically focused infrastructure interspersed throughout the urban realm, in all its manifestations and scales, that is connected spatially and socially to the communities that it serves, offering a suite of outputs and opportunities for engagement.
- **Environmental emancipation:** pertaining to the right to freely experience and benefit from accessing nature and biophilic places and the wider qualities of the environment without penalty and realise the positivity of the linked health, educational, social and related benefits.

The Urban Environment as a Socio-Ecological System

The idea of the urban landscape being viewed as a socio-ecological system has been discussed in some detail in earlier chapters. Its use as a term equates well with the concept of biophilia and the power of the human-nature bond, but also links well concerning the diversity of places, the different typologies and spatial characteristics found in urban when considering their contribution to people and the quality of the environment (Mutnick et al., 2022).

It is advantageous to describe the urban environment as an ecosystem, i.e. as a functioning dynamic system, but the disciplinary remit of ecology is so closely linked to the study of natural systems and nature conservation that reference to ecology and ecological value in urban situations can involve unhelpful comparisons with natural systems. While natural systems are arguably a proper point of reference, it has had the effect of downgrading spaces that do not demonstrate the qualities of true natural ecosystems, even if they provide important social outputs, a situation that clearly is unacceptable. However, embracing terms such as socio-ecological, especially in an urban context, moves to recognising and promoting space and green space characteristics that have high social outputs, as well as those which are recognised primarily for their natural characteristics and species diversity. Accordingly, while there is no need to reject existing descriptions of places or values, this approach does embrace a wider analysis and recognition of places regarding their potential socio-ecological performance, and helps develop appropriate strategies concerning the ability of such places to provide multi-outputs and connect to local communities. In fact, such an approach merely acknowledges current thinking regarding ecosystem performance, and helps to overcome any existing prejudice regarding different environmental typologies and their ecological attributes.

The socio-ecological system is of course a dynamic one, both in terms of the relationship between the two defining elements, but also through the processes of nature, seasonal and longer temporal change, that sees vegetation systems mature, but also design and development implications whereby places change use, or interventions that modify the previous trajectory and which often sees the social or ecological significance increase or be re-assessed in the light of observed function or fuller understanding. All places have some form of social or/and ecological outputs, even if they are not well recognised as a specific part of their identity – although, of course, many are set aside for just these purposes receiving designation and protection, although others show shockingly weak recognition.

The system exists as a basic dynamic irrespective of artificial boundaries including ownership patterns. It embraces all levels of interaction and involvement, which encompasses such aspects as walking through, taking care of, looking at, meeting in, conservation, the protection and celebration of nature; the urban landscape as a socio-ecological complex works well within the context of biophilia and how the city is read. By examining and understanding the performance and potential performance of the environment (the social and ecological outputs), it is possible to bring infrastructure more explicitly into the socio-ecological system, as well as enhancing those places that are already recognised, plus identifying new niches of operation. At its most fundamental level, it concerns appreciating the significance of green infrastructure, the balance between built and natural forms and the connections that are made. This includes the biophilic credentials assessed at both local and city-wide levels and the parameters of the green spaces as critical, rather than discretionary, parts of urban systems.

Under this remit, new niches of socio-ecological interaction are being realised all the time as understanding of the system becomes more sophisticated and, for example, are demonstrated throughout the case studies (Chapter 9). The socio-ecological system recognises that nature is a powerful driver when it comes to realising multiple ambitions, but the adjustment of terminology does recognise and emphasise our current understanding of human-environment relationships as a more universal quality, and that nature and human endeavour exists as a complementary and synergistic relationship rather than separate considerations.

The Choreography of the Landscape Experience

As discussed in Chapter 4, the idea of choreography within a landscape context suggests that there is an element of performance and direction regarding the ways in which the environment is experienced. Indeed, this is so, since the individual reacts to different prompts and cues initiated by different landscape forms, assesses the opportunities and relevance to their lifestyles and responds accordingly. Such a concept fits easily within the discipline of landscape design, but it also embraces the different theoretical strands regarding landscape perception, urban design, the image of the city and biophilia. This also includes the ways that biophilic associations are experienced in practice, and can be explored in the ways that nature (and thus biophilic associations) needs to be interspersed throughout urban areas in providing a balanced environment and the most effective way that this can be achieved.

The idea of choreography can be expressed in a number of ways, but concerns a resonance between user and the environment involving the stimulation of the senses which comes to the fore when person and place are in tune (even though they may not be aware of it). The notion of choreography takes account of the interaction and the flow that occurs when immersed in a landscape situation, and how the individual negotiates this flow by their actions, for example the pace of movement, speeding up, slowing down, looking at, listening to, stopping, moving towards or the haptic connections involved in, for example, cultivating the soil (Csikszentmihalyi, 2002).

Choreography, in the context used here, thus refers to the bodily immersion and the rhythm of walking, style and circumstances of engagement, including the multi-modal sensory experience that takes place – how the senses pick up and interpret these external cues. It is influenced by the innate biophilic response, but also determined by cultural and social contexts, including past experiences, memories and personal associations. Choreography is a summation of the process which is directed by individual experience. The associated biophilic linkages can be described by personal place dynamics, including the socio-ecological context in which they are set, relaxation, chilling out, a preferred meeting place, promoting healthy exercise, conservation work, watching nature, urban agriculture and growing food, as well as social interaction rooted in place identity. It is also clear from the studies reported on that certain pre-requisites are necessary to achieve an effective linkage of factors. This includes how an effective result might be demonstrated, e.g. in relieving mental stress or providing a feeling of being away (Chapter 2), or may be identifiable as a broader social discourse and community identity. These are aspects that are well articulated in the various strands of literature, also becoming recognised in standards of provision, including those regarding access to and the distribution of urban green space or in precedent examples which demonstrate the same qualities (Chapter 9).

The idea of choreography links well to theories of landscape perception and also those that are culturally based, as the individual reacts to an environmental encounter experiencing changes in the rhythm and flow of the encounter. Biophilic focused reactions are consistently demonstrated as a major part of a positive force and clearly provide a draw, where it is clear that the experience is dependent on the strength or resilience of the landscape. This concerns the extent to which the vegetation and its (biophilic) properties provide a buffer against unwanted intrusions, but also, how previous encounters provide the necessary reference points for interpretation. Places and environments are interpreted in ways that are meaningful to the individual and meaningful to the senses, and relate to positive personal associations and linked memories of places, people, activities or events, but are set in the context of an innate biophilic response.

The context of these findings is diverse and suggests that the idea of choreography can be applied across the spectrum of urban environments or sub-environments and across theoretical frameworks, including interpretation in different cultural contexts, involving different life stages, although there are specific sub-criteria for the most effective delivery. It is clear that memories of past places are used to interpret new encounters; in pleasurable environmental encounters, moments are captured, and remembered, to be used in evaluating the affordances of new situations. Linked to an innate biophilic response, the various personal and cultural factors combine, or indeed fail to combine. When they do, they form a positive feedback loop where the memories, current experience and the properties of a new encounter are compatible or congruent (Chapter 1). They can then form cherished moments of sensory and emotional satisfaction. It is these cherished moments that are re-visited when appropriate environmental associations cue their response. Thus, it seems likely the development of environmental awareness and the characteristics of a personal choreography are fine-tuned by repeated pleasurable exposure to the environment, but that it also builds upon early childhood experiences, and this is the situation discussed in Chapter 6. It is also clear that, especially in biophilic situations, experiences are underpinned by the intensity of sensory feedback (sight, sound, touch) to give a 'buzz' or peak flow of emotional response (Csikszentmihalyi, 2002), which is evident in the ways that people respond (Chapter 4).

A rich environment is thus related to how effective it is in relation to 'natural' movement and activity set with personal identity and experience. It is also clear that a number of strategies and other interventions can support and heighten the experience and are demonstrated across situations; for example, community-based initiatives (Chapter 9), supporting environmental literacy (Chapter 8) or sensitising the individual by art-based interventions or related self-administered techniques (Chapter 5). However, in the absence of appropriate environments or opportunity to experience them, none of this will happen. Where they do happen, a positive response is indicated, but the personal rationale underpinning that response will be exactly that, synthesised in personal and cultural terms.

Different people may have the same broad positive reaction to an environment or place, but their specific reasons for doing so may be quite radically different. In essence, this means that individuals can interpret the environment according to their own personal remits (including cultural experience) without any problem, and will most likely 'see' and relish the features or outputs that are congruent with their own experiences and expectations. This means that the development of the urban environment and urban green infrastructure, as informed by biophilic principles and biophilic design, can comfortably be pursued as an all-influencing and pervasive aspect of our urban centres with the formation of a naturally focused environment capable of multiple personal interpretations and accommodating a range of activities.

Choreographies contain an emotional component, and appear to be driven by a desire for personal satisfaction and the reaffirming of personal identity, but also appear to be supported in the adoption of a mindful approach, being aware of one's own responses and feelings, which in turn, enables one to more fully understand the experience. For example, in our own work, when invited through self-narration, our volunteers found it surprisingly easy to articulate their feelings, and in doing so revealed the choreography of their encounters and took us into their personal experiences and emotional states, thus allowing us to understand, from their perspective, the processes and context that made their experiences meaningful, more powerful and long lasting (Chapter 4). However, choreographies can also be considered in terms of place dynamics and circumstances, where powerful eco-social relationships can be established in new circumstances by adopting strong biophilic solutions – for example, in the design of the workplace or the functioning of streets (Chapter 9). These involve new choreographies which start to redefine and expand our understanding of socio-ecological circumstances and, in turn, influences how we consider the function and functionality of buildings and streets. However, it also changes the dynamic surrounding work-based social interaction, performance and how things could be done. These are situations that suggest that the creation of biophilic environments can accommodate individual choreographies and situation to considerable advantage (Beatley, 2011: 2009).

A Continuously Connected Biophilic Landscape System

The delivery of an effective functional urban environment, in terms of spatial organisation and quality, is underpinned by the idea of a connected system, connected in both the way that different parts link together, but also how they connect to the populations they serve and the ability of people to make connections both physically and socially, i.e. that they have to confidence to experience them. Or, put another way, it is the potential of the environment to facilitate connections (health, social, community, educational) that is determined by its accessibility and quality. However, from the earlier discussion it is evident that connection can be considered more widely as a fundamental property of a socio-ecological system.

Achieving connection can be addressed under the remit of biophilic design, i.e. embracing nature in all its forms, and as a focus for action, is implicit in understanding people, place and wider city relationships (Kellert, & Calabrese, 2015; Tebb, 2020). There are outstanding examples of how nature is being embraced, respected and delivered in new urban developments (Chapter 9) which demonstrate both what could be and what can be done, as new niches of opportunity are realised and partnerships formulated. Similarly, there are city-scale initiatives, especially through the Biophilic City movement (biophiliccities.org), which demonstrate biophilic credentials as a central ideology; again, they include strong visions and strong delivery. In addition, there are new niches being exploited, including green roofs, terrace gardens and eco-pod approaches, which facilitate biophilic connections under very different circumstances (Chapter 9) and further define the remit of the socio-ecological system and invite further exploration of all situations and sectors.

In this context, connections can also be related to pathways. The basic remit is that nature acts as a pathway, and that places which demonstrate strong biophilic qualities also demonstrate strong attributes in relation to, for example, pathways to health and well-being, or community and local identity. This can be seen at different levels, from nature in its capacity as a moderating influence on urban form, through to being immersed in the authentic properties of wilder places. Pathways and connections can thus be understood from different perspectives, including different disciplinary starting points, allowing pathways and connections to be realised and reinforced in respect of different objectives. This includes the operational feature of the pathway, but also allows a certain amount of relaxation regarding how things can be done, in that connections and potential pathways can be recognised and considered as both physical reality and operationally, delivering multiple objectives and multiple outputs. Precedents and current knowledge demonstrate the ways forward and are becoming increasingly refined in their approach as the ways that new pathways and connections are identified and are increasingly underpinned by a biophilically focused environmental philosophy.

These bring new socio-ecological resources online, where they feature more explicitly in the socio-ecological system, and there is no reason why the same approach cannot be adopted over an entire city administration, in particular utilising partnerships, identifying new opportunities, re-assessing existing place dynamics and responding accordingly. This concerns all disciplines and sectors and calls for a similar degree of innovation to underpin the provision of nature places throughout. However, it is at the local scale where connections are likely to be addressed in the most subtle of ways, but also where there is substantial opportunity and need, again involving partnerships, making connections, understanding how residents use their environment, identifying potential resources and realising strategies through a combination of nature, eco-focused place initiatives and supporting social activities. The significance of small-scale interventions, as well as the potential resources that they can bring in terms of both people and places, is a situation that offers scope for additional innovation.

Environmental Emancipation

On health grounds alone, there is a case for ensuring that everyone can access the health promoting aspects of walking in high-quality 'green', and it can be argued that there is a natural justice to ensure equitable access to green infrastructure and ensure that this includes opportunities to experience nature in the form of wild spaces within easy walking distance from where people live (Chapter 3). Promoting, protecting, managing and accessing urban places in ways that realise their full potential responds to the same socio-ecological parameters that describe its functioning. While there are certain sectors and professions which have particular skill sets to manage and develop urban green, emancipation also involves the wider role of society in supporting developing and managing the environment that goes hand in hand with its use.

Environmental emancipation involves both basic rights regarding access and provision, but also requires the adoption of supportive attitudes, taking responsibility towards the environment and addresses all sectors. This also requires that people have the necessary skills to do so, through an appropriate level of knowledge and environmental literacy. This applies equally to the professional remit and that professionals are appropriately educated (are environmentally literate) or that children are supported in their acquisition of environmental skills (Chapter 6).

Understanding the functioning of urban green infrastructure is likely to lead a more equitable situation and produce a more powerful and resilient system. However, this is not the only consideration; it is necessary to take into account the human factor, including the personal barriers and the lethargy involved in environmental amnesia (Chapter 8). Promoting environmental literacy, giving people the confidence to access nature and realise its benefits by improving environmental literacy and skills, including a greater awareness of its influence and what can be achieved are important aspects and can be seen within the profiles of biophilic cities (biophiliccities.org).

Emancipation involves responsibilities and rewards, an understanding of nature and the natural world that leads to stewardship, involving a wish to help protect and manage the resource. Within this context, as well as improving and providing additional green space, there is a mobilisation of resources in terms of citizen power involving community ventures, green-space champions, educational and support mechanisms, events and celebration, which support people in gaining the confidence they need to access and enjoy the power of nature, but also moving to protect and manage that resource in the form of stewardship. Indeed, it can be argued that such aspects as practical conservation work, supporting others, leading nature walks and celebrating successes are an integral part of the socio-ecological system in which human endeavour and ecological stewardship are component parts of the equation.

The current agenda concerns ideas of co-production, including professional/public partnerships and local authority/other sector ventures, which involve the development and implementation of a shared agenda (Box 10.1) but also

BOX 10.1 SOME PARAMETERS RELATING TO ENVIRONMENTAL EMANCIPATION DERIVED FROM THE CASE STUDIES AND RELATED EXAMPLES

- **Birmingham City of Nature Plan** (Chapter 9) – challenging the status quo, 25-year inclusive vision, people focused, environmental justice tool, nature recovery charter involving community council partnerships, environmental charter, standards for urban parks.
- **Forest School** (Chapters 6 & 9) – embedding environmental literacy, stewardship and actions and confidence for access for later life; linked social and educational development.
- **Gardens for Wildlife Victoria** (Chapter 9) – environmental literacy, supporting actions, training and education, shared values and common purpose, dissemination of ideas, local authority/citizen partnerships, realising the potential of citizen power and gardens as a resource, upgrading the ecological quality of spaces, realising social and community benefits, publicity and community of action.
- **Warndon Villages** (Chapter 9) – environment as a recognised resource, embedding local identity through the natural landscape and heritage features, realising and releasing the power of nature, letting nature take the lead, a supporting network of green corridors for walking and cycling, strong biophilic identity, local environment plan, strong actions supported by the parish council, and residents, local green champions and advocates and linked initiatives, conservation work, support ventures including walking.
- **Queen Elizabeth Hospital** (Chapter 3) – realising the potential of private space, public use of the hospital ground as a growing space and health resource for staff and patients, voluntary sector/National Health Service partnership, bringing new spaces space into the socio-ecological system, utilising the power and pleasure of food growing and horticultural activities as a basis for healthy actions.
- **Admiral Kampung Singapore** (Chapter 9) – a new vision for residential space, integrating nature and living space in a balanced way application of biophilic principles in strong social settings, green roofs and terraces and a community farm, set within the context of Singapore as a biophilic city.

combining resources which may include expertise, workforce and financial aid. Co-production is inherent in environmental emancipation in defining and facilitating the pursuit of a common agenda, responsibilities and actions being particularly effective when pursued in a holistic manner involving local strategies and actions that are integral to the wishes of the communities they serve, but also within the wider context of city administration (Mumaw & Raymond, 2021). In particular, the community knows its local environment and has the capacity to undertake biophilic and other environmental audits, provide an enthusiastic workforce, undertake surveys, develop supporting activities such as wardening, leading walks and organising events, where local champions may emerge. These are activities which benefit from the production of local plans or in the celebration of achievements, and opportunities do exist to develop local environmental plans as part of statutory legislation which enshrines local ideals in a statutory planning framework.

Conclusions – New Urban Landscapes, New Opportunities and Attitudes

In putting material together and interrogating case studies, a clear momentum towards the idea of what constitutes a healthy city and the involved person-place dynamics is demonstrated. It is possible to appreciate the range of theoretical and practice-based perspectives in the reality of modern living and to realise life strategies that embrace health, address societal and planetary needs and foster environmental awareness, including care and stewardship of the environment.

The component pieces are there, as can be seen in case studies and by interrogating experience, both in personal and professional capacities, and how this addresses all facets of urban life. The initiatives and findings involved reflect a current trend, in regard to the urban landscape, that seeks to re-connect society to nature and natural processes, and is seen in changing attitudes to the landscape – allowing nature to take the lead. As a future trajectory for the urban landscape, it is clear that society is questioning how we care for the environment, and there is desire to allow nature back into our everyday lives, a situation which is being explored and pursued by professional and individuals with ideas, examples and precedents at all levels.

The identification of the environment as a socio-ecological system is thus as much about attitudes and responsibilities as action, and how these play out regarding landscape and social initiatives in the real urbanised world. The urban aesthetic being pursued and recommended owes its basis to the understanding of and embracing the power of nature in all sorts of ways that are reflected in the sustainable city and the biophilic city. As the Biophilic City website observes, all cities are already biophilic to some degree, and thus it is a case of building upon these foundations. Understanding how we react to the urban environment and promoting its structure and quality accordingly moves society to a situation where the positive effects are likely to become increasingly evident, opportunities to realise them universal and the advantages to personal health, planetary health, social cohesion and emotional and personal well-being more profound, providing that nature and its power are allowed to fully re-claim their place in an urbanised society.

BOX 10.2 THE DEVELOPMENT OF THE URBAN ENVIRONMENT AS A SOCIO-ECOLOGICAL RESOURCE

- Key aspects of social and eco form a continuum, underpinning the understanding of and functioning of the urban environment which respects and embraces the importance of nature as an essential component in urban lifestyles.
- Embrace the power of nature in a socio-ecological context, having regard to how the city is read and how it displays its environmental credentials.
- Recognise the importance of different environmental characteristics and associated place dynamics in social as well as ecological terms.
- Understand the importance of the environment and accessing nature regarding health and in the pursuit of healthy lifestyles.
- Learn from precedents at all scales, ownership and development patterns.
- Recognise and celebrate biophilic situations and achievements at all levels of spatial organisation and ownership, note the precedents of Biophilic Cities, local studies and examples of professional practice.
- Recognise the importance and functioning of biophilic places across the spectrum of the socio-ecological continuum and audit their potential.
- Take an eco-centric approach, view infrastructure as a linked system, overcome issues relating to different ownership, bring actions together under a common purpose.
- Develop partnership, take a co-production approach.
- Address environmental emancipation and environmental justice to ensure that populations are freely able to access their environment and assume responsibility for it.
- Pursue biophilic audits and personal actions, especially related to health, well-being, and stewardship.
- Activate people power as a resource, including measures to realise latent biophilic responses.
- Ensure environmental literacy across generations, including within school education programmes.
- Ensure an appropriate distribution of and access to wild nature and related green spaces throughout urban centres to at least accord with recommendations for distribution, access and the ideals of environmental justice; where appropriate, undertake measures to enhance their natural biophilic characteristics.
- Progress the agenda of an urban aesthetic that embraces the power of nature, seeking to improve the connectivity of the socio-ecological system; address actions at all scales.
- Embrace web-based communication platforms as part of the socio-ecological system to celebrate people's individual endeavours, and maintain contact with the natural world.

- Address the importance of contact with nature and the effects of biophilia in diverse situations and activities across all sectors, work, play, recreational and residential.

The extent to which urban socio-ecological systems are new thinking or simply reassert an ancient biophilic heritage and its exploration in a modern form as an urban aesthetic can be argued about, but it is clear that such an aesthetic is bounded by clear credentials, validated by much research and practical examples which concern health, social and ecological remits. The indications are that biophilic connections are part of the human condition, and the implications are that they need to be pursued as an integrated part of urban functioning. It is in the context of urbanisation and modern urban lifestyles that we find ourselves exploring ideas of nature and environment, and in which we and others have been re-examining experience and applying ideas within academic, educational and professional practice contexts. The place of nature in the urban world is found to be both present and needed, being expressed in many different forms and relationships, although it should be remembered and recognised that the presence of nearby wild places still constitutes an essential part of modern human experience.

It is everyday life events, played out in urban environments, that represent the reality of much of the population, and it is within the reality of urban lifestyles that these relationships need to be explored. Interrogating the various strands of literature reveals that this is essentially about how the body and mind are located in urban space. Explored in theory and applied in practice, the state of knowledge is substantial, and continues to provide examples and impetus for action, including personal and professional responsibilities, direct action and broader educational and supportive interventions.

References

Beatley, T. (2009). Biophilic urbanism: Inviting nature back into our communities and into our lives. *William & Mary Environmental Law and Policy Review* 34(1), 6. https://scholarship.law.wm.edu/wmelpr/vol34/iss1/6. Available at researchgate.net

Beatley, T. (2011). *Biophilic Cities: Integrating Nature into Urban Design and Planning*. Washington, DC: Island Press.

Csikszentmihalyi, M. (2002). *Flow: The Psychology of Optimal Experience*. New York: Random House.

Frank, B., Delano, D. L., Laniglia, B. S. (2017). Urban systems: A socio-ecological system perspective, *Sociology International Journal* 1(1). www.researchgate.net/publication/317369842

Kellert, S., Calabrese, E. (2015). *The Practice of Biophilic Design*. www.biophilic-design.com

Mumaw, L., Raymond, C. M. (2021). A framework for catalyzing the rapid scaling of urban biodiversity stewardship programs, *Journal of Environmental Management (Epublication 15 August 2021)* 292(18), 112745.

Mutnick, D., Cuonzo, M., Griffiths, C., Leslie, T., Shuttleworth, J. M. (2022). *The City is an Ecosystem: Sustainable Education, Policy, and Practice*. London: Routledge.

Tebb, P. J. (2020). *Biophilic Urbanism: Designing Resilient Communities for the Future*. Abingdon: Routledge.

INDEX

Note: Page numbers in *italics* indicate a figure on the corresponding page.

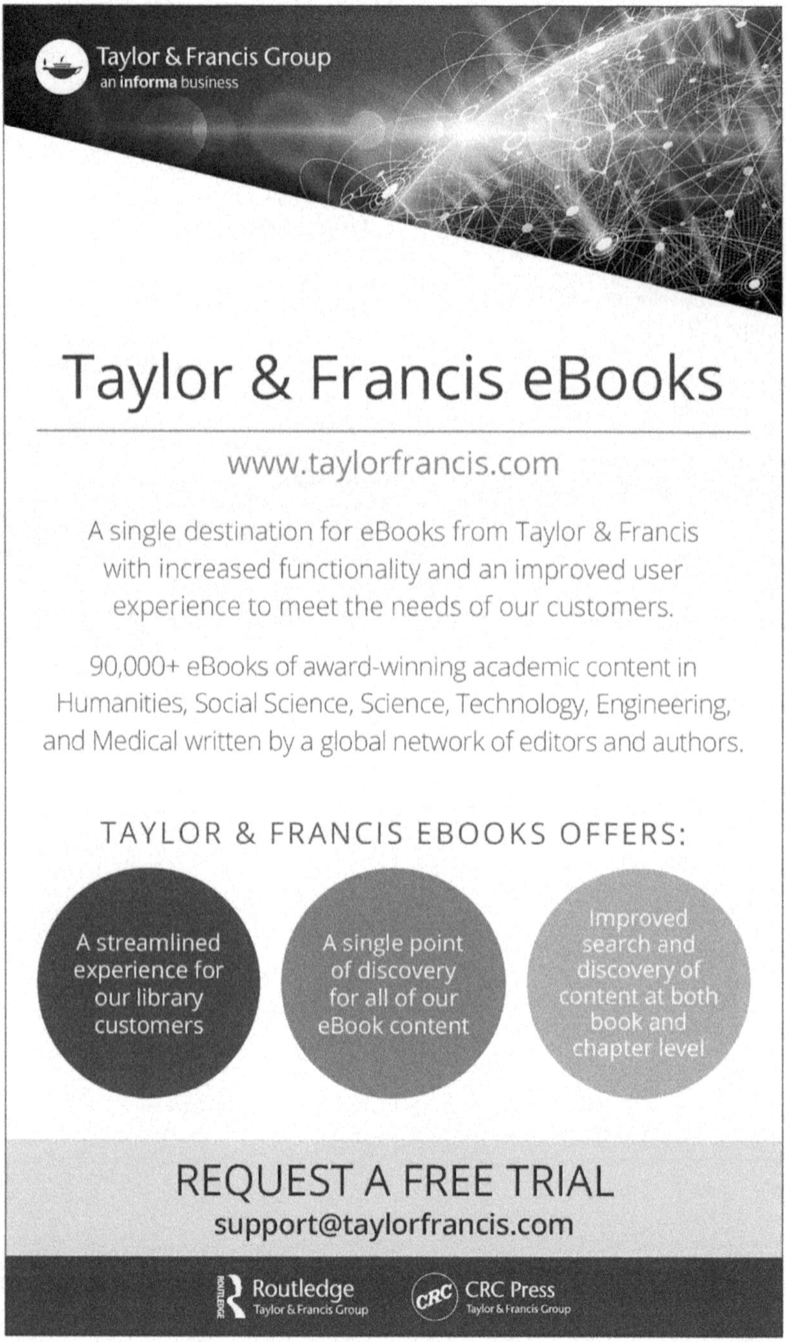

Taylor & Francis Group
an informa business
Taylor & Francis eBooks
www.taylorfrancis.com
A single destination for eBooks from Taylor & Francis with increased functionality and an improved user experience to meet the needs of our customers.
90,000+ eBooks of award-winning academic content in Humanities, Social Science, Science, Technology, Engineering, and Medical written by a global network of editors and authors.
TAYLOR & FRANCIS EBOOKS OFFERS:
A streamlined experience for our library customers
A single point of discovery for all of our eBook content
Improved search and discovery of content at both book and chapter level
REQUEST A FREE TRIAL
support@taylorfrancis.com
Routledge
Taylor & Francis Group
CRC Press
Taylor & Francis Group

For Product Safety Concerns and Information please contact our EU representative GPSR@taylorandfrancis.com
Taylor & Francis Verlag GmbH, Kaufingerstraße 24, 80331 München, Germany

www.ingramcontent.com/pod-product-compliance
Ingram Content Group UK Ltd.
Pitfield, Milton Keynes, MK11 3LW, UK
UKHW020150200926
475702UK00016B/2640
9780367568740